TEMPORAL LOGIC

From Philosophy and Proof Theory
to Artificial Intelligence and
Quantum Computing

TEMPORAL LOGIC

From Philosophy and Proof Theory
to Artificial Intelligence and
Quantum Computing

Stefania Centrone
Klaus Mainzer
Technical University of Munich, Germany

World Scientific

EW JERSEY · LONDON · SINGAPORE · BEIJING · SHANGHAI · HONG KONG · TAIPEI · CHENNAI · TOKYO

Published by

World Scientific Publishing Co. Pte. Ltd.

5 Toh Tuck Link, Singapore 596224

USA office: 27 Warren Street, Suite 401-402, Hackensack, NJ 07601

UK office: 57 Shelton Street, Covent Garden, London WC2H 9HE

Library of Congress Control Number: 2023006260

British Library Cataloguing-in-Publication Data
A catalogue record for this book is available from the British Library.

TEMPORAL LOGIC
From Philosophy and Proof Theory to Artificial Intelligence and Quantum Computing

ISBN 978-981-126-853-3 (hardcover)
ISBN 978-981-126-854-0 (ebook for institutions)
ISBN 978-981-126-855-7 (ebook for individuals)

For any available supplementary material, please visit
https://www.worldscientific.com/worldscibooks/10.1142/13205#t=suppl

Desk Editors: Logeshwaran Arumugam/Rok Ting Tan

Typeset by Stallion Press
Email: enquiries@stallionpress.com

Preface

Calculi of temporal logic are widely used in modern computer science. The temporal organization of information flowing in the different architectures of laptops, the Internet, or supercomputers would not be possible without appropriate temporal calculi. But the situation is similar to modern artificial intelligence (AI) and other digital technologies. In the age of digitalization and high-tech applications, people are often not aware or have forgotten that temporal logic is deeply rooted in the philosophy of modalities, which dates back to antique philosophers, such as Aristotle and Theophrast.

At the dawn of the modern scientific era, Leibniz and others came up with the first ideas of formal calculi and machine-based decision procedures, which culminated in the proof theory of the 20th century, connected with names such as Hilbert, Gentzen, Gödel, and Turing. The debates on mathematical foundations at that time underline the important role of constructive and intuitionistic procedures for proof theory, which are also convenient for computational algorithms in computer science. Becker and Gödel discovered a link between intuitionistic and modal logic. Automatic reasoning in modern AI and robotics would not be possible without the constructive methods of proof theory and mathematical logic.

Therefore, a first goal of this book is to become aware of these roots in philosophy and proof theory (Section 1.1). A deep understanding of these roots opens avenues to the modern calculi of temporal logic, which have emerged by the extension of modal logic by temporal operators (Section 1.2). Chapter 2 is dedicated to the computational foundations of temporal logic in different calculi with

v

increasing complexity, such as basic modal logic (BML) (Section 2.1), linear-time temporal logic (LTL) (Section 2.2), computation tree logic (CTL) (Section 2.3), and full computation tree logic (CTL*) (Section 2.4).

Chapter 3 highlights the proof-theoretical foundations of temporal logics. A fundamental role for proof-theoretical interpretations of the temporal calculi is played by the sequent calculus of Gentzen (Section 3.1). The tableau-based calculus (Section 3.2), automata-based calculus (Section 3.3), game-based calculus (Section 3.4), and dialogue-based calculus (Section 3.5) are more or less inspired by a Gentzen-style sequent calculus. They are applied as proof-theoretical interpretations to the calculi of the modal and temporal logics, which were introduced in Chapter 2. It is remarkable that these different interpretations of temporal logics have different advantages but also different disadvantages for different purposes, especially in computer science. Therefore, it makes sense to have these different approaches of proof theory to temporal logics.

Chapter 4 provides an outlook on trend-setting applications of temporal logics in future technologies, such as AI and quantum technology. Modern machine learning is based on statistical learning theory and stochastic procedures, which need control policies for safety and security. At this point, formal tools of temporal calculi come in to certify AI programs, which are characterized by complex information flow in machine learning (Section 4.1).

In quantum technology, quantum physics and computer science are growing together. Traditional temporal logics assume classical physics and its concept of time as self-evident. But we have to consider the different concepts of time in the different physical theories. They have been studied in different temporal logics of, for example, relativistic physics (Section 4.2). For quantum technologies, modal and temporal operators significantly depend on basic concepts such as quantum parallelism in quantum computers and entanglement in quantum communication (Section 4.3). Temporal logics are obviously deeply embedded in current digital technologies. Therefore, the chapter concludes with an outlook on the societal impact of temporal logic, which is needed to handle the increasing complexity in a computational, high-tech world of Big Data (Section 4.4).

The book is written by the two authors with respect to their earlier and ongoing studies and complementary competence. Stefania Centrone wrote Sections 1.2 and 3.1. Klaus Mainzer wrote Sections 1.1, 2.1–2.4, 3.2–3.5, and 4.1–4.4.

About the Authors

Klaus Mainzer is Emeritus of Excellence at the Technical University of Munich (TUM) and senior professor at the Carl Friedrich von Weizsäcker Center of the University of Tübingen. After studies of mathematics, physics, and philosophy and a Heisenberg award at the University of Münster, he was a professor of the foundations and history of exact sciences and vice president at the University of Constance, a professor of philosophy of science and the director of the Institute of Interdisciplinary Informatics at the University of Augsburg, and a professor of philosophy of science, the director of the Carl von Linde Academy, and the founding director of the Munich Center for Technology in Society (MCTS) at TUM. His principal research interests are in the constructive and computational foundations of mathematics, science, and philosophy, with a special focus on AI technology and its societal impact.

Stefania Centrone is a Professor of Logic and Philosophy of Science at the Technical University of Munich (TUM). She studied philosophy at the University of Florence from 1994 to 1999 and got her PhD at the Scuola Normale Superiore, a university institution of higher education based in Pisa. In 2012, she got the German habilitation in philosophy at the University of Hamburg, where she also received her *venia legendi* in 2013. She holds the Italian habilitation for ordinary professorship in the logic, philosophy, and history of science. With the award of the Heisenberg position, Stefania Centrone has been recognized as belonging to the top 5% of researchers of excellence in Germany, according to the German Research Foundation (DFG).

Contents

Chapter 1

Philosophical Roots
of Temporal Logic

1.1. The Concept of Time

Pre-Socratic natural philosophers, such as Parmenides and Heraclitus, formulated basic questions of time that influence the discussion even today. Is the world, as Heraclitus believed,[1] in constant becoming and time an irreversible process like the flow of a river? Or is every change, as Parmenides believed, only apparent and time a reversible parameter of an inherently unchanging world?

Zeno's famous paradox of the arrow of time illustrates the problem: "If everything that behaves in a uniform manner is either in constant rest or constant motion, but everything that moves is always in the now, then the flying arrow is without motion." Thus, during the duration of a moment, an arrow in motion takes a route from which it does not move away during that moment. Also, in the next moment, it covers this distance, from which it does not move away during this moment. But how can it then move away at all, however small the distance between two moments may be?[2]

Mathematically, the criticism of Zeno's paradox is simple if (as in modern physics) the flight distance of the arrow is understood as a real number continuum. Then, there is simply no "next" moment

---

[1]Diels-Kranz (1960/1961).
[2]Ferber (1981).

because the points in the continuum are dense, so between two adjacent points, an intervening point can always be given (e.g., by halving the distance between them).

Democritus' atomic theory can be understood as a consequence of the Heraclitean doctrine of change and the Parmenidean principle of unchangeable being. While objects such as stones, plants, and animals are, according to Democritus, aggregates of atoms, which can change in time and combine into new atomic groupings, atoms and empty space are timeless, uncreated, and eternal. The task of physics, according to Aristotle, is to explain the principles and functions of diversity and change in nature.

1.1.1. *Aristotelean temporal logic*

The general principle that makes an individual being, such as a stone, a plant, or an animal, what it is, Aristotle calls the *form*. That which is determined by the form is called *matter*. Form and matter, however, do not exist by themselves but are principles of nature obtained by abstractions. Matter is therefore also called the *possibility* ("potency") of being formed. Only by the fact that matter is formed, *reality* arises. *Movement* is determined as *change*, as a transition from possibility to reality, or as "actualization of potency." According to Aristotle, movement includes all goal-oriented processes in nature, such as the fall of a stone to the Earth, the growth of a tree from seed to final form, and the development of man from infant to adult.

This offers the following solution to Zeno's paradox of the arrow of time. In doing so, Aristotle criticizes Zeno's definition of the present. The present, Aristotle argues, is no more part of time than a point is part of a distance. Rather, one must think of the present as a potential, non-actual cut in the time continuum that separates the future from the past. In that case, the present is not a point in time in which the arrow is actual but only a possibility. In fact, the arrow performs a continuous movement.

Aristotle is the first philosopher to formulate the concept of continuum precisely. Time, he says, is continuously connected in the now. But time has no existence of its own. Real are only the movements of nature. The now of a moment is a cut in the continuum of movement. Since one can potentially make an unlimited number of cuts in the continuum, countably many moments result without ever being able to exhaust the continuum.

"Time," says Aristotle, "is not, however, motion, but what is countable about it."[3] As a measure of time, Aristotle proposes circular motion, against which other motions can be measured. The assumed spherical motions of the stars and planets are for him a reference system to measure, for example, hours, days, and years. The uniform circular motion is considered the basic measure of ancient and medieval astronomy.

Against the background of his philosophy of time and continuum, Aristotle develops a logic of temporal modalities. Real is what is now realized (i.e., true) at the moment. Possible is what is realized (i.e., true) at a present or future time. Necessary is what is realized at the present and in any future time. Thus, the now-relativization is characteristic in his definition of modalities.

While Stoic logic subscribes to this definition of the modalities of time, the Megaric school of philosophy uses the modalities of possibility and necessity even without now-relativization. Possible is what is realized at a certain time. Necessary is what is realized at any time.

The Aristotelian analysis of the sentence "Tomorrow a sea battle will take place" became famous.[4] It is already true that tomorrow a sea battle will take place or that tomorrow a sea battle will not take place. Does it not include that either it is already true that a naval battle will take place tomorrow, or that it is already true that a naval battle will not take place tomorrow? Does it not include, then, that whichever of the two statements (of which we do not yet know which) will be true tomorrow is already true today?

In the symbolism of temporal logic, p denotes a state, such as "a naval battle is taking place." The symbol Np is supposed to mean that the state denoted by p will take place tomorrow. As a time measure, for example, at noon, the highest position of the sun can be indicated. The symbol $N\neg p$ then means that the negation $\neg p$ of the state denoted by p takes place tomorrow, i.e., p does not take place. The symbol $\vee$ stands for the logical or-connection of two statements describing two states. Then, the symbol $N(p \vee \neg p)$ says

---

[3] Aristotle (1956, p. 219b).
[4] Aristotle (1994, pp. 27–97).

that it is already true that the state described by p will exist or not exist tomorrow. But does it follow that, according to the Aristotelian problem, $Np \vee N\neg p$, i.e., is it already true that the state denoted by p will exist tomorrow or that its negation $\neg p$ will exist tomorrow?

1.1.2. *Linear and branching temporal logic*

Modern logicians, such as G. H. von Wright, have shown that the answer to this question depends on the topology of time evolution on which we base the world. In a linear conception of time, the time segments (e.g., days) are arranged sequentially on a straight line. Each node in Fig. 1.1(a) stands for the overall state (i.e., a conjunction of many partial states) of the world at a given time interval. The filled node ($\bullet$) stands for the now updated overall state. The subsequent nodes denote future overall states, and the previous ones denote past overall states. In this time model, the subsequent states are uniquely determined without alternative. Therefore, in this world, it is true now that a naval battle will take place tomorrow or that no naval battle will take place tomorrow.

But if we assume the possibility of several future development branches, we get the picture of a time tree as in Fig. 1.1(b), where the actualized now ($\bullet$) is followed by several possible overall states in the next temporal period (e.g., days), which again can be followed by several possible overall states in the next but one temporal period, etc. The dotted lines indicate the possibility of a future development branch. The dotted lines also indicate that in past time periods, several possibilities may have existed, which were not realized, however. The past is therefore a linearly arranged sequence of total states to the following time periods.

The assertion that $N(p \vee \neg p)$ is true at a given temporal period holds in both development models. At the present time interval ($\bullet$), in both Figs. 1.1(a) and 1.1(b), it is true that the partial state, denoted by p, either is or is not a component of any overall state following in the next time interval (*tertium non datur*). However, the assertion that $Np \vee N\neg p$ is true at a given time interval is generally true only in the linear world (Fig. 1.1(a)). Indeed, in the case of a branched future world (Fig. 1.1(b)), it would have to be true at the present time interval that the partial state denoted by p is part of any subsequent

(a)

(b)

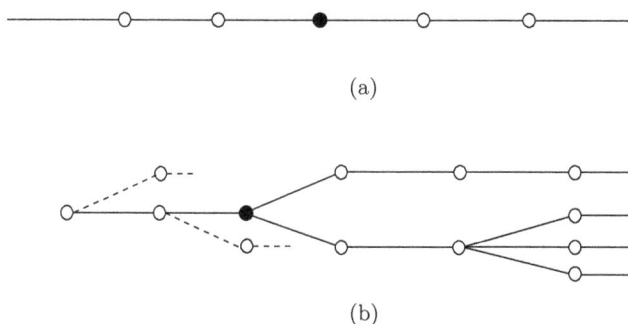

Fig. 1.1.   (a) Linear time. (b) Branching tree of time.

overall state in the next time interval (Np), or it would have to be true that the partial state denoted by p does not belong to any subsequent overall state $(N\neg p)$.[5]

In the example of an ancient sea battle, the condition of the already present truth Np is probably never satisfied. However, one can easily think of conditions where $N\neg p$ is already true today. For example, all ships are distributed in the world today in such a way that they could not possibly be gathered into a fleet tomorrow. If $N\neg p$ is true, then the truth of $Np \lor N\neg p$ also follows.

Linear and branching time are relevant for several arguments and positions in the history of philosophy. According to William of Ockham (1288–1347), propositions about the future have a definite truth value, although only God can know them. But Ockham did not support determinism or fatalism that denies human free will. The question arises whether the branching-time model needs to assume a privileged branch.[6] Obviously, Ockham rejected the Aristotelian idea that, in order to preserve the contingency of the future, future contingents must be deemed neither true nor false. Later, C. S. Peirce (1839–1914) criticized the assumption that future contingents can have definite truth values. For Peirce, only the present and past are actualized. The future is open to possibility and necessity.

---

[5] von Wright (1974, pp. 161–178).
[6] Santelli (2022).

1.1.3.   *Temporal logic of modern physics*

The debate on Aristotelian temporal logic illustrated that concepts of temporal modalities often depend on physical assumptions. In modern physics, the Ptolemaic planetary model with its uniform circular motions is no longer a reference system for time measurement. Celestial bodies are understood as reference systems in which clocks measure time. To the question "What is time?" Einstein had answered succinctly, "Time is what a clock measures." According to this, celestial bodies in the Universe are moving reference systems with space and time coordinates, in which lengths are measured with scales and time with clocks. Newton still assumed the "resting" Universe as absolute space with absolute time, with which all moving clocks can be synchronized.

According to Einstein's special theory of relativity, time measurement is no longer absolute but becomes path dependent. Every moving body therefore has its own relativistic time ("proper time"). Thus, there is no absolute distinction between the past, present, and future but only relative to the observer who has his or her proper time. This relativistic insight is reflected in a famous quotation of Einstein which he mentioned at the occasion of the death of his Swiss friend Michele Besso in March, 1955:

> Now he has also preceded me a little with his farewell to this strange world. That means nothing. For us believers in physics, the distinction between past, present and future has only the significance of an illusion, albeit a stubborn one.

As a physicist, Einstein does not mean a subjective time of experience but a metrically, topologically, and objectively precise concept of time. For St. Augustine, the distinction of past, present, and future is only in the mind ("soul") of an observer:

> But this much is now clear and distinct: neither the future nor the past "is," and not actually can be said: times "are" three: past, present, and future; ... For it is these times as a kind of trinity in the soul, and elsewhere I do not see them; and there is the presence of the past, namely memory; the presence of the present, namely sight; the presence of the future, namely

expectation. If we are allowed to speak in this way, then I also see three times and admit: yes, there are three.[7]

In general relativity, spatio-temporal reference systems that are under the influence of gravitational forces are taken into account. This causes space-time to be "curved." In Section 3.1, it is shown what changes this results in for temporal modalities in the temporal logic.

On the basis of the general theory of relativity, the cosmological standard models for the possible developments of the Universe can be derived mathematically. The differences between linear- and branched-time logic with time trees must be taken into account. Some models allow finite and infinite temporal evolutions with an initial singularity ("Big Bang") and final singularity ("Big Crush"). In the Christian tradition of philosophy, an initial beginning of time ("Creation") and an ending of time ("Day of Judgment") were already assumed (e.g., by St. Augustine). Black holes, whose existence follows from general relativity, are also physically time singularities in which time evolutions (trajectories) end. In the dawning age of space travel, time travel has become interesting and is theoretically possible in some models. Particularly noteworthy is Gödel's solution to Einstein's gravitational equation, which allows for circular time trajectories.

For a physical decision on these time models, however, quantum mechanics must also be taken into account. According to Heisenberg's uncertainty principle, time measurement depends on Planck's quantum of action: The more precisely the time of a quantum system (e.g., an elementary particle) is measured, the more the measured value of its energy scatters and vice versa. Time and energy are conjugate quantities in quantum mechanics. The path of an elementary particle from an initial location to a future location is no longer determined by a unique time trajectory (line). Rather, according to Feynman's path integral, all possible connecting paths between the two locations must be taken into account ("summed up").

---

[7] Augustine (2016, Book XI).

Classical physics, relativity, and quantum mechanics, however, leave untouched an important aspect of time that is essential to our everyday understanding of time. We experience time as irreversible: We are born, age, and die. The reversal has never been observed. In contrast, the basic laws of classical physics, relativity, and quantum mechanics are time-invariant, i.e., they also allow for the reversible time transformation $t \to -t$. In thermodynamics, however, irreversible processes are studied: Milk poured into coffee mixes into latte. The reversal has never been observed. Traditional temporal logic often assumes an everyday understanding of time with irreversible development that would first have to be physically justified by statistical mechanics and thermodynamics. In that case, as shown in Section 3.1, time can no longer be understood as a parameter (*parameter time*) but as an operator that describes irreversible entropy processes of complex statistical systems near and far away from thermal equilibrium (*operator time*).

1.2.   Toward Formal Systems of Temporal Logic

From the late 19th to the early 20th century, new concepts of temporal logic were developed against the background of new philosophical tendencies and emerging formal logic. Therefore, on the historical path to formal systems of temporal logic, we start with the remarkable, but less considered, philosophical and formal studies of the mathematician and philosopher Oskar Becker in 1930. In an essay titled "On the Logic of Modalities," he states[8]:

> Considered from the standpoint of modalities [...] the problem of *mathematics* and *temporality* [...] receives a strong, though unilateral, clarification. [...] The "logic" of *modalities* has a deep relation to temporality.

This section sets out to formally explore the relation between modalities and temporality based on Becker's considerations.

---

[8]Becker (1930, p. 43, op.). Henceforth, page numbers of references to Becker (1930) are from the original pagination.

Let us first state that temporal propositions contain some reference to time conditions. Classical logic consists of timeless propositions, such as:

A: "The sun is a star."
B: "The sun is rising."
C: "The sun is setting."

Proposition A is timeless since it is true in the past, present, and future. Propositions B and C refer to the time condition "now." While formulas of classical logic refer to static states and properties, formulas of temporal logic describe sequences of state changes.

A temporal logic results from an extension of a classical propositional or predicate logic through temporal operators which introduce temporalized modalities.[9] There are *at least* four temporal operators, **G**, **H**, **F**, and **P**:

— **G** (Guarantee): **G**A: the proposition that A will be always true.
— **H** (History): **H**A: the proposition that A was always true.
— **F** (Future): **F**A: the proposition that A will be true.
— **P** (Past): **P**A: the proposition that A has been true.

The operators "**G**" and "**H**" denote *necessity* both in the past and the future. The operators "**F**" and "**P**" denote *possibility* both in the past and the future. For the sake of simplicity, we denote them by the operators normally used in the current logic of alethic modalities: *necessity* ($\Box$) and *possibility* ($\Diamond$), with an index "p" or "f" to refer to the past or the future, respectively. Thus:

— **G** (Guarantee): $\Box_f A$: the proposition that A will be always true.
— **H** (History): $\Box_p A$: the proposition that A was always true.
— **F** (Future): $\Diamond_f A$: the proposition that A will be true.
— **P** (Past): $\Diamond_p A$: the proposition that A has been true.

In principle, classical logic can express temporal properties too but through complicated formulas with quantifiers relating to points of time.

---

[9]See van Benthem (1983).

Linear time refers to sequences of states. If linearity of time is assumed, additional operators, such as "next" and "until," can be introduced. In this case, each state has exactly one future. In branching-time logic, a state can have several futures that refer to a branching tree of states. Tree models are used to model nondeterministic developments.

Before coming to a formal system of temporal logic, we analyze Becker's text against the background of the main philosophical questions about the nature and the properties of time[10]: Does time have a beginning or an end? Is time linear and directional, branching, or circular? Does Becker comprehend time as instant-based or interval-based? Does he take time to be discrete, dense, or continuous? Does our perception of time depend on the structure of consciousness? Does it depend on physical measurement? It turns out that Becker's analysis is inspired by mathematical intuitionism. The final question is: How do we best choose a formal language that is suitable to express Becker's ideas about time? The formal systems in question are even remarkable for applications in linguistics, computer science, and artificial intelligence.

1.2.1.   *Modo recto and modo obliquo*

"One can [...] say" — so says Becker in "On the Logic of Modalities" — "that the philosophical 'discovery' of the '*modi obliqui*,' *possibility* and *necessity*, which have acquired their ontological acknowledgement only after the '*modus rectus*,' the '*truth*' (reality) (and its opposite, the 'falsity' or 'unreality'), represent a first step towards the discovery of the 'authentic' temporality. For, the characteristic 'modi' of the latter, the *future* and the '*past*,' can be literally designated as '*modi obliqui temporales*' as opposed to the temporal '*modo recto*,' that is, the '*present*.'"[11]

Let us first note that the passage just quoted says that (i) there is an authentic temporality, even if Becker doesn't say what it is, that (ii) the alethic modalities, *possibility* and *necessity*, pave the way to discover such an authentic temporality, and that (iii) the characteristic *modi* of the latter are the *future* and the *past*.

---

[10]Mainzer (2002).
[11]Becker (1930, p. 43).

What about *modo recto/modo obliquo*? *Modo recto/modo obliquo* is a terminology that, as readers familiar with the Austrian-Polish tradition would know, is often associated with the name of Franz Brentano. Roughly, one represents something *modo recto* if one refers directly to it and *modo obliquo* if it is a secondary object of one's intention. So, if one thinks of Dr. Bernard Bolzano's edifying speeches to the academic youth, one represents, according to Brentano, the edifying speeches *modo recto* and the academic youth *modo obliquo*. We will later confront Brentano's use of the expressions *modo recto/modo obliquo* at a variance with Becker's. For the moment, we just state that they are different and focus on the latter.

According to Becker, the expressions *modus rectus — obliquus* come from Latin grammar and correspond to the *verbal moods* rather than to the *verbal tenses*: The *indicative* is the "direct" *mood*, the *subjunctive* is the "indirect" (or "oblique") *mood*. The *indicative* corresponds to the *actuality* and the *subjunctive* to the *non-actuality* of a process that takes place over time. Becker singles out, *prima facie*, three *modalities*: contingency (*recto*), possibility and necessity (*obliquo*), and three temporal dimensions: present, past, and future. He writes[12]:

> The expressions "modus rectus — obliquus" [...] are [...] modelled [...] on the names of the "verbal modes" [...]. The "tempora" of the verb, not its modi, seem to denote linguistically the temporal modalities. However, the "tempora" only denote the non-analysed or already constituted temporality, that is a temporality that has already been subsumed under the schema of an objective process (event). The origin of time, the "enactment-character [*vollzugsmäßige Weise*]" of its peculiar "temporalizing itself [Sich-Zeitigens]" is rather rendered through the *modi verbi*.

Thus, let us state that Becker's considerations are guided by the use of *verbal moods* and *not* of *verbal tenses* in natural language, at a variance with most temporal logics, as, for instance, Prior's logics, which are rather inspired by the use of tenses in natural language. Becker seems to take the verbal modes to be the syntactic expression of the conceptual pair *realis — irrealis*: The *indicative* is seen as the syntactic expression of the *realis* and the *subjunctive* as the syntactic

---

[12]Becker (1930, p. 43, fn.).

expression of the *irrealis*. In her book, *The Languages of Native North America*, Marianne Mithun characterizes the difference as follows: "The *realis* portrays situations as actualized, as having occurred or actually occurring, knowable through direct perception. The *irrealis* portrays situations as purely within the realm of thought, knowable only through imagination."[13]

Summing up, according to Becker, the alethic modalities of *possibility* and *necessity* have an inner connection with the *temporal* modalities of *future* and *past* as opposed to the *present*. The *future* and the *past* are marked as *modi obliqui temporales*, and the present is marked as *modo recto*. *Obliqui* alludes to the fact that both the *future* and *past* are *non-factual* or *non-actual*. For instance, what is in the future has not yet occurred and can therefore only be in the realm of possibility. But also, what is located in the past has already occurred and is therefore non-actual. It seems to be a common trait of natural languages to render the opposition between *realis* and *irrealis* by means of the distinction between indicative and subjunctive.

1.2.2.   *Intuitionism and the relationship of modalities and temporality*

Becker writes: "The origin of time [...] is rather rendered through the modi verbi." When he talks of time this way, Becker refers to time as a structure of the mind.

"Time" — so says Kant in his *Critique of Pure Reason* — "is not an empirical concept that is somehow drawn from experience."[14] We would not perceive simultaneity or succession if our consciousness would not structure what reaches our sense organs through time. Only if the sense of time is given *a priori*, before any experience, it is possible to represent several things existing at one and the same time (simultaneously) or in different times (successively). Kant conceives time as a necessary *a priori* structure of our consciousness. This means that we continuously organize what we experience through time.

---

[13]Mithun (1999, p. 173).
[14]Kant (1787, p. B46).

Kant assumed that human understanding is made possible by two forms of intuition, which are given to the subject before (*a priori*) any empirical experience (1787):

(1) Human subjects have an intuition of spatial forms which are constructed step by step through certain rules (e.g., a circle and ruler in geometry).
(2) Human subjects have an intuition of sequential temporal points which are constructed step by step by adding a unit according to the rule of counting in arithmetic.

On the one hand, the spatial and temporal forms of intuition enable the human subject to understand empirical objects and events in space and time. For Kant, human subjects are equipped *a priori* with an intuitive understanding of space and time, which makes possible their orientation in the world. On the other hand, the spatial and temporal forms of intuition provide the schemes of geometric and arithmetic constructions and, by that, the foundations of mathematics.

Kant illustrates the temporal form of intuition by an unlimited sequence of points:

$$., ..., \cdots, \cdots,$$

which is extended step by step by a unique point. In our temporal intuition, these points represent a sequence of present moments ("now") which are passing in a linear order. Formally, this process corresponds to the construction of the natural numbers $1, 1+1, 1+1+1, \ldots$ or in the usual abbreviation of decimal numbers $1, 2, 3, \ldots$ according to the rule of counting.

After Kant, the mathematician Leopold Kronecker (1823–1891) claimed that natural numbers are "made by God," but "all the rest" is made by humans (according to Weber, 1893).[15] Independent of Kronecker's reference to God, Kant argues on the same line: Natural numbers are given to the subject by the temporal form of intuition and its arithmetical scheme of counting. "All the rest" must be reduced to the fundamental form of arithmetical construction.

---

[15] *Die ganzen Zahlen hat der liebe Gott gemacht, alles andere ist Menschenwerk.* (Weber, 1893, p. 19).

Thus, for Kant and Kronecker, infinity is not given but only a *façon de parler* (Poincaré) or "regular idea" (Kant) for the unlimited process of counting.

According to Brouwer, mathematical truth is founded by constructions of a "creative subject."[16] In his intuitionistic foundation of mathematics, he followed Kant's explanation of human understanding by human subjects. Brouwer derived radical consequences for the truth of theorems and their proofs: In temporal intuition, only finite sequences of natural numbers can be constructed.[17] Thus, for Brouwer, mathematical truth depends on finite stages of realization in time by a creative subject. Later, Georg Kreisel suggested the following formal definition.[18]

A creative subject has a proof of proposition A at stage m (in short: $\sum \vdash_m A$) iff

(CS1) For any proposition A, $\sum \vdash_m A$ is a decidable function of A, i.e.,

$$\forall x \in \mathbb{N} \left(\sum \vdash_x A \vee \neg \sum \vdash_x A \right);$$

(CS2) $\forall x, y \in \mathbb{N}(\sum \vdash_x A \to (\sum \vdash_{x+y} A))$;
(CS3) $\exists x \in \mathbb{N}(\sum \vdash_x A) \leftrightarrow A$.

The idea that only finite initial segments of infinite sequences are given leads to Brouwer's concept of choice sequences.[19] A choice sequence means a process which is not necessarily predetermined by some law or algorithm:[20]

(i) α *lawless sequence* $:\equiv$ at any stage of $\alpha 0, \alpha 1, \alpha 2, \ldots$, only finitely many values of α are known.
(ii) α *lawlike sequence* $:\equiv$ all values of α are known by a law (i.e., algorithm).

Lawless sequences can be illustrated by sequences of casts with a die after some already realized casts in the beginning. The die can be

---

[16]Brouwer (1907).
[17]See Mainzer (2018, Chapter 6).
[18]See Kreisel (1967), Sundholm (2014).
[19]See Mainzer (1977).
[20]See Troelstra (1968), Dummett (1977).

thrown in arbitrarily many times. Nevertheless, at any stage, only finitely many casts are known, and the following cast is unknown. In lawlike sequences, all stages are predetermined by a law. An example is the sequence of even natural numbers with law $\alpha n = 2n$ or the sequence of decimal places of real number π.

A radical departure from classical mathematics occurred when choice sequences were allowed in real analysis. If real numbers are defined by fundamental sequences to be given by choice sequences, then the statement "all total functions from $\mathbb{R}$ to $\mathbb{R}$ are continuous" can be proven intuitionistically. At first glance, this statement seems to be false in classical mathematics. But the meaning of the condition "function f from $\mathbb{R}$ to $\mathbb{R}$ is total" is obviously much stronger if $\mathbb{R}$ is extended to real numbers defined via fundamental sequences as choice sequences[21] (Brouwer, 1927). Therefore, we get a different intuitionistic meaning of classical concepts.

In intuitionistic mathematics, infinite objects are considered as ever growing and never finished. Therefore, sets need a new foundation. The intuitionistic analog of a set is a so-called spread, which is defined as a countably branching tree labeled with natural numbers or other finite objects and containing only infinite paths. A fan is a finitely branching spread. A branch is an intuitionistic choice sequence, i.e., an infinite sequence of numbers (or finite objects) created step by step by a law (algorithm) or without a law (e.g., coin). A lawless sequence is ever unfinished. The only available information about a lawless sequence at any stage is the initial segment of the sequence created thus far.

Here, temporal logic comes in because, for Brouwer as well as for Becker, time sequences and sequences of numbers are closely connected: "The reason why number and time belong together is obvious: The infinite series of numbers is — by its infinity — determined by the potentiality and with that by futureness. Again, the special character of the 'obliquitas' is decisive."[22]

Thus, following Brouwer and Becker, future is realized in finite stages, such as intuitionistic spreads, i.e., countably branching trees with infinite paths. Only the initial parts of the trees are given as

---

[21] Brouwer (1927).
[22] Becker (1930, p. 46).

past and presence. Intuitionistically, continuity of time is essential. Concerning the spreads of branching paths in the future, they are realized in finite steps. Becker states in the end of his essay[23]: "The idea of a path contains finity [...] the most abstract and 'most free' sciences of mathesis universalis are primarily and principally determined by the concept of finity."

1.2.3. Becker's modal-logical interpretation of intuitionism

In his 1930 essay, "On the Logic of Modalities," Becker was the first logician and mathematician to put forward the idea of a *modal interpretation* of intuitionistic logic, more precisely, the idea of a possible sound and faithful translation of intuitionistic logic into modal logic. However, the first actual translation is to be found in a one-page celebrated and influential paper entitled "An interpretation of the intuitionistic propositional calculus" written in 1933 by Kurt Gödel.[24] The basic idea of Gödel is similar to the one Becker outlines in his essay. Becker suggests adding to classical logic the predicates "(...) is provable," "(...) is such, that its negation is provable," and "(...) is undecided." Such predicates should express Brouwer's primitive logical concepts.

Similarly, Gödel's idea is to add to the language of classical propositional logic the unary operator "it is provable that (...)," denoted by "B," and to an axiomatic calculus for classical propositional logic *three* axioms and *one* rule of inference. The axioms correspond exactly to the modal schemas K, T, and 4[25] that characterize modal logics that are nowadays standard, and the rule of inference is the necessitation rule that is contained in all *normal* modal systems.

Note, incidentally, that both Becker and Gödel seem to take the predicate "(...) is provable" and the operator "it is provable that (...)" as conveying the same piece of information. Actually, the predicate "(...) is provable" denotes the property of a proposition to be provable, while the operator "it is provable that (...)" takes a

---

[23]Becker (1930, p. 51).
[24]Gödel (1933).
[25]For a formal definition of the normal modal logics and an explanation of their characterizing axioms see Chapter 3.

proposition as input and gives a different proposition as output. (Unfortunately, this practice of systematically neglecting the difference between predicate and operator is, even nowadays, quite widespread among logicians.)

Gödel writes[26]:

> One can interpret Heyting's propositional calculus by means of the notions of the ordinary propositional calculus and the notion "p is provable" (written "Bp"), if one adopts for that notion the following system S of axioms:
>
> 1. $Bp \to p$
>
>   if it is provable that p, then it is true that p
>
> 2. $Bp \to ((B(p \to q) \to Bq)$
>
>   if it is provable that p and it is provable that p implies q, then it is provable that q
>
> 3. $Bp \to BBp$
>
>   if it is provable that p, then it is provable that it is provable that p
>
> In addition, [...] the new rule of inference is to be added
>
> $$\frac{A}{BA}$$
>
> From A, it is provable that A may be inferred

By substituting throughout the operator "B" ("it is provable that $(\ldots)$") by the operator "$\square$" ("it is necessary that $(\ldots)$"), one obtains one of the modal logical systems that are nowadays standard, namely Lewis's system S4.

Becker did also tentatively consider a number of possible ways to extend Lewis's S3 in order to get a modal system with a finite number of irreducible modalities, or even an infinite number of irreducible modalities, yet all *pairwise comparable* with respect to logical strength. In the current context, the most interesting of Becker's "experiments" is the one he discusses at length in Section 5, Part I, of his essay.[27] Becker tackles here in a more abstract way the problem of "completing" Lewis's calculus in such a way that in the resulting

---

[26] Gödel (1933, p. 301).
[27] Becker (1930, pp. 25–30).

system, let us call it SM,[28] independently from the number (finite or infinite) of irreducible modalities, any two (positive) modalities (i.e., combinations of the two elementary modalities $\Box$ and $\Diamond$) be comparable with respect to logical strength:

> One can try to free oneself from all requirements that are imposed by special *material* assumptions, and introduce solely the most general *formal* conditions of a calculus of modalities. Firstly, one can drop the requirement that the reduction to a finite number of primitive modalities be possible. [...]
>
> One will, however, keep the general requirement of a linear rank order of modalities, whereby the implicational relation of any two distinct modalities is univocally determined. Otherwise, a modality could no longer be univocally determined by its "rank of logical strength." Such requirement should be in any case the upper bound of our formal freedom.
>
> Now, the question is, on the one hand, whether Lewis's calculus [S3] satisfies this requirement and, on the other hand, whether suitable axioms for a theory of the rank order of modalities can be established independently from Lewis's calculus. The answer to the first question is negative, while the answer to the second question is positive.[29]

At the end of a rather elaborate argument, he arrives at the claim that a stepwise generalization of Brouwer's schema:

$$B_1 = \Box(A \to \Box\Diamond A),$$
$$B_2 = \Box(A \to \Box\Box\Diamond A),$$
$$B_3 = \Box(A \to \Box\Box\Box\Diamond A),$$
$$\vdots$$

provides an infinite number of axiomatic conditions, which added to a given base modal calculus (which he does not explicitly specify) together with three meta-rules $(R_1) - (R_3)$,[30] produce a calculus SM whose modalities, although infinite in number, are linearly ordered.

---

[28]Centrone and Minari (2019, p. 54).
[29]Becker (1930, p. 25).
[30]These rules impose that the logical strength relations between (composed) modalities be suitably preserved under "multiplication" of modalities.

He leaves as an open problem the question whether all these infinite modalities of SM are irreducible.

Concerning SM, Becker's proof that the (supposedly infinite) modalities of this system are linearly ordered is very clever and correct. Unfortunately, he did not notice that, *if the base calculus is Lewis's* S3, already the first two schemas, B_1 and B_2, of the infinite sequence of schemas $\{B_n\}_{n\geq 0}$ he postulated, together with the rule R_2, are sufficient to prove the schema $\Box(\Box A \rightarrow \Box\Diamond A)$ — and thus to make SM equivalent to the normal modal system S5.

However, it can be proved[31] that by choosing a suitable (semantically characterized) base modal system weaker than S2 (and incomparable with S1), Becker's "system SM" does not collapse into S5 while having the properties that Becker expected.

1.2.4. *Intuitionism and formal systems of temporal logic*

There is a rich variety of models of time: Is time instant-based or interval-based, or is it discrete, dense, or continuous? Does time have a beginning or an end? Is time linear, branching,[32] or circular? These models must satisfy appropriate formal systems of temporal logic. The flow of time is defined by a model $\mathcal{T} = \langle T, \prec \rangle$ with a (nonempty) set T of time instants and a binary relation $\prec$ of precedence.[33] In a discrete (forward, resp. backward) model, each time instant which has a successor, resp. predecessor, also has an immediate successor, resp. predecessor. In dense models, there is another instant between any two subsequent time instants. In an instant-based model of time $\mathcal{T} = \langle T, \prec \rangle$, many properties can be formulated as first-order sentences, such as reflexivity, irreflexivity, transitivity, asymmetry, linearity, beginning, end, no beginning, no end, density, and discreteness.

Second-order sentences with quantification over sets are required for properties, such as continuity and well-ordering. Here, the different meaning of classical and intuitionistic sets as spreads comes in.

---

[31]See Centrone and Minari (2022).
[32]See Belnap (1992).
[33]Galton (2008, p. 3).

Obviously, Becker argues for an intuitionistic semantics of temporal logic with spreading temporal developments in finite steps.

If events with duration are under consideration, it is sometimes more convenient to use interval-based models instead of instant-based models of time.[34] Interval-based models avoid the paradox of Zeno's flying arrow: If at each instant the flying arrow stands still, how is movement possible? An example that also requires interval-based reasoning is: "The housebreaker was robbing the house, when the police came." Interval-based models are richer than instant-based models of time because, besides temporal precedence $\prec$ between time instants, they also consider inclusion and overlapping of time intervals. From a mathematical point of view, instants can be considered as infinitesimal small intervals.

The first formal system TL of temporal logic was introduced by Prior,[35] who was motivated by the use of tense in natural languages. That reminds us of Becker's analysis of *modo recto* and *modo obliquo*. In TL, the propositional language with atomic propositions and classical logical connectives is extended by Prior with four temporal operators:

P: "It was the case that ... " (with P for "past"),

F: "It will be the case that ..." (with F for "future"),

G: "It always will be the case that ...,"

H: "It always was the case that ...."

The operators G and H can be defined by P and F and vice versa with

$$G :\equiv \neg P \neg \qquad H :\equiv \neg F \neg.$$

If π is an infinite path in a spread, then these operators can be combined, for example:

$\pi \models FG\varphi$: "At a certain instant, φ is true at all future states of the path,"

$\pi \models GF\varphi$: "φ is true at infinitely many states on the path."

---

[34] See Allen and Hayes (1989).
[35] See Prior (1957).

Thus, the syntax of TL is specified with the grammar

$\varphi, \psi := a|\bot|\neg\varphi|\varphi \vee \psi|P\varphi|F\varphi$ with atomic formula a.

A model is given by $\mathcal{T} = \langle T, \prec, V \rangle$ with a frame (Kripke) $\langle T, \prec \rangle$ and valuation function V which assigns a truth value to each pair (a, t) of an atomic formula and a time value. $\mathcal{T} \models \varphi[t]$ means that φ is true in a model $\mathcal{T} = \langle T, \prec, V \rangle$ at time t. The following formal system is called the minimal temporal logic K_t.[36] Axioms:

φ (with φ tautology of first-order logic)

$\mathrm{G}(\varphi \to \psi) \to (\mathrm{G}\varphi \to \mathrm{G}\psi)$,

$\mathrm{H}(\varphi \to \psi) \to (\mathrm{H}\varphi \to \mathrm{H}\psi)$,

$\varphi \to \mathrm{GP}\varphi$,

$\varphi \to \mathrm{HF}\varphi$,

with rules of deduction

$$\frac{\varphi \to \psi\varphi}{\psi} \quad (modus\ ponens),$$

$$\frac{\varphi}{\mathrm{G}\varphi} \quad \text{(with } \varphi \text{ tautology)},$$

$$\frac{\varphi}{\mathrm{H}\varphi} \quad \text{(with } \varphi \text{ tautology)},$$

An example which can be derived is Becker's rule:

$$\frac{\varphi \to \psi}{\mathrm{T}\varphi \to \mathrm{T}\psi}$$

with a tense T for any sequence of operators $\mathrm{G, H, F}$, and P. A translation from statements of TL into first-order logic with one free variable t representing the present instant of time can be defined recursively.

---

[36]Rescher and Urquart (1971, Chapter VI).

Additional axioms can be taken to represent further assumptions of time. Examples[37] are

Transitivity (TRANS):	$G\varphi \rightarrow GG\varphi$, $H\varphi \rightarrow HH\varphi$, $FF\varphi \rightarrow F\varphi$, or $PP\varphi \rightarrow P\varphi$;
Reflexivity (REF):	$G\varphi \rightarrow \varphi$, $H\varphi \rightarrow \varphi$, $\varphi \rightarrow F\varphi$, or $\varphi \rightarrow P\varphi$;
Linearity (LIN):	$(PF\varphi \vee FP\varphi) \rightarrow (P\varphi \vee \varphi \vee F\varphi)$;
Well-ordering with transitivity (WELLORD):	$H(H\varphi \rightarrow \varphi) \rightarrow H\varphi$;
Time without end (NOEND):	$F\top$ or $G\varphi \rightarrow F\varphi$;
Induction (IND):	$F\varphi \wedge G(\varphi \rightarrow F\varphi) \rightarrow GF\varphi$.

Minimal temporal logic K_t can now be extended to systems of temporal logic, for example:

$K4_t := K_t + TRANS$: all transitive frames,

$S4_t := K_t + REF + TRANS$: all partial orderings,

$L_t := K_t + TRANS + LIN$: all strict linear orderings,

$N_t := L_t + NOEND + IND + WELLORD$: $\langle \mathbb{N}, < \rangle$ natural numbers.

These temporal logic systems are decidable.[38]

Over discrete and linear models of time, the basic temporal logic can be extended by a *Next Time* operator X and the operators *Since* S and *Until* U. $X\varphi$ is true at an instant of time t iff φ is true at the immediate successor $s(t)$ of t. The *Next Time* operator satisfies the axioms

$$X(\varphi \rightarrow \psi) \rightarrow (X\varphi \rightarrow X\psi) \quad \text{and}$$

$$X\neg\varphi \leftrightarrow \neg X\varphi.$$

The binary operator $\varphi S\psi$ means that φ has been the case since a time when ψ was the case. $\varphi U\psi$ means that φ will be the case until a time when ψ is the case. The extension of TL with these operators lead to the linear-time temporal logic (LTL). LTL is interpreted over

---

[37] Galton (2008, p. 10).
[38] Rescher and Urquart (1971), Burgess and Gurevich (1985).

the structure of natural numbers with a reflexive temporal ordering. Obviously, LTL is appropriate for Becker's intuitionistic assumptions of time. Practically, LTL is convenient to express assertions on infinite computations in computer science.

From an intuitionistic point of view, linear time only refers to one path in a branching spread which represents alternative possibilities for the future. A spread is graphically represented by a branching tree. Openness of future means intuitionistically that the temporal structure $\langle T, \prec \rangle$ is never given as a closed infinity but growing in finite steps: "The idea of a path contains finity [...]."[39]

It turns out that in a branching time, two alternative semantics are possible for the future operator F, which were denoted by Prior as Peircean and Ockhamist interpretations.[40] In the Peircean branching-time logic, the future operator F means that "it will necessarily be the case that ...." Consequently, future truth is assumed as truth in all possible futures. But with this understanding, the future operator F can no longer be dual to the strong future operator G: Fφ is true at an instant t iff every temporal development passing through t contains some later instant t' at which φ is true. The strong future operator G does not change its meaning ("It will necessarily always be the case ..."). Therefore, in the Peircean version, G must be supplemented to the formal language as additional primitive operator and cannot be reduced to F.

In Peircean temporal logic, the formula $\varphi \to$ HFφ is no longer valid. From an intuitionistic point of view, the principle of excluded middle is crucial. The Peircean interpretation preserves the principle of excluded middle as $\varphi \lor \neg\varphi$, but it violates the principle of future excluded middle F$\varphi \lor$ F$\neg\varphi$. This principle means that, eventually, either φ or $\neg\varphi$ will be the case.

In the alternative Ockhamist interpretation, truth is not only relativized to a time instant in the temporal tree but also to a temporal development ("history") passing through this instant. The operator F now means "With respect to the given history, it will be the case that ...." Therefore, an Ockhamist tree model is a triple $\mathcal{T} = \langle T, \prec, V$ with tree $\langle T, \prec \rangle$ and valuation function V which assigns a truth value

---

[39]Becker (1930, p. 51).
[40]Prior (1967).

to each triple (a, t, h) of an atomic formula a, a time value t, and a history h. Then, $\mathcal{T}, t, h \models \varphi[t]$ means that φ is true in a model $\mathcal{T} = \langle T, \prec, V \rangle$ at time t and history h.

All temporal formulae which are valid in linear models of time are also true at all instant-history pairs, i.e., valid in the Ockhamist sense. Therefore, the Ockhamist interpretation confirms the principle of future excluded middle $F\varphi \vee F\neg\varphi$ if time does not end. It also validates the formula $\varphi \rightarrow HF\varphi$. But the principle of the necessity of the past $P\varphi \rightarrow \square \, P\varphi$ is violated. Decidable and strongly complete axiomatizations of intuitionistic temporal logic were recently proven.[41]

1.2.5. _Outlook on temporal logic in computer science_

Logical systems of branching time are not only interesting philosophically but also for computer science.[42] Peircean and Ockhamist branching-time temporal logics correspond to the computational tree logics CTL[43] and CTL*.[44] In computer science, branching time is understood as a computation tree. In this case, histories have the order type of natural numbers, such as in intuitionistic mathematics. In computer science, trees of data are unfolded by discrete transition systems which generate possible infinite computations.

In general, a transition system is defined by actions or processes which realize transitions between system states. The infinite behavior of transition systems can be expressed in terms of paths and computation. Transition systems are ubiquitous in computer science and can be specified as finite-state automata, pushdown automata, Turing machines, etc.[45] Practically, all kinds of computers or clocks can be considered as transition systems. Thus, transition systems deliver an operational semantics of temporal logic.

The computation tree logic CTL* follows the Ockhamist interpretation of temporal logic. The reflexive future operators G and U and the _Next Time_ operator X are interpreted on computation trees.

---

[41]Boudou _et al._ (2017), Chopoghloo and Monini (2021).
[42]See Gabbay _et al._ (1994, 2000), Bolc and Szalas (1995).
[43]Emerson and Clarke (1982).
[44]Emerson and Halpern (1985).
[45]Demri _et al._ (2016, Chapter 3).

An evaluation is relativized to an instant of time defined as state of a transition system and a history defined as a computation path. The computation tree logic CTL follows the Peircean interpretation of temporal logic. Each of the temporal operators G, U, and X is immediately preceded by a modal operator □ or ◊.

Philosophically, transition systems overcome the restriction of temporal logic to the temporal structure of human consciousness. A constructive and intuitionistic semantics of temporal logic is interesting for cognitive systems with their conscious structure of time in the tradition of Kant and Husserl, and Becker and Brouwer. But it can also be realized by machines in computer science and artificial intelligence.[46] For example, in computer science, the correct behavior of a reactive program with nonterminating computations requests to specify and verify the acceptable infinite executions of the program. In concurrent programs with processors working in parallel, their interaction and synchronization must also be specified and verified.[47] An infinite computation is formulized by the linear-time logic LTL. Nondeterministic systems are modeled in a branching-time logic. Besides LTL, CTL and CTL* are important logical tools to specify and verify reactive and concurrent computational systems.

In artificial intelligence, nowadays, machine learning[48] applications need temporal organization of computational processes. AI-supported translation systems need an operational semantics of tenses in natural languages. Cognitive robots (e.g., autonomous cars) with self-organizing abilities need a kind of spatial–temporal reasoning. All that was, of course, not foreseen by Becker as future applications of temporal logic. But his constructive and intuitionistic approach opens a future perspective for an operational semantics of temporal logic.

---

[46]Ohrstrom and Hasle (1995), Fischer *et al.* (2005).
[47]Pnueli (1977, pp. 46–57).
[48]Mainzer (2019, Chapter 11.1).

Chapter 2

Computational Foundations of Temporal Logic

In Section 1.2, we already mentioned transition systems (TSs) as general models of computational systems. They consist of states which represent certain configurations and transitions which represent possible state changes. Changes are determined by actions. In this way, finite state machines, automata, pushdown systems, Turing machines, counter machines, and supercomputers can be modeled. But real-life systems can also be modeled by TSs, such as clocks, smartphones, vending machines, and robots. All kinds of computational architectures, such as sequential, parallel, reactive, and interactive processes (e.g., in robotics), can be realized by TSs. Geometrically, they can be illustrated as directed graphs with labels on vertices or edges, which allow software verification by model checking.

Definition of transition systems:[1]

A routed and interpreted transition system (ITS) is a structure

$$T = \left(S, \left\{ \xrightarrow{a} \right\}_{a \in Act}, L, s \right)$$

consisting of

(1) a set $S \neq \emptyset$ of states,
(2) a set Act of names (labels) of actions acting on states and producing successor states,

---

[1]Demri *et al.* (2016, Section 3.1.1).

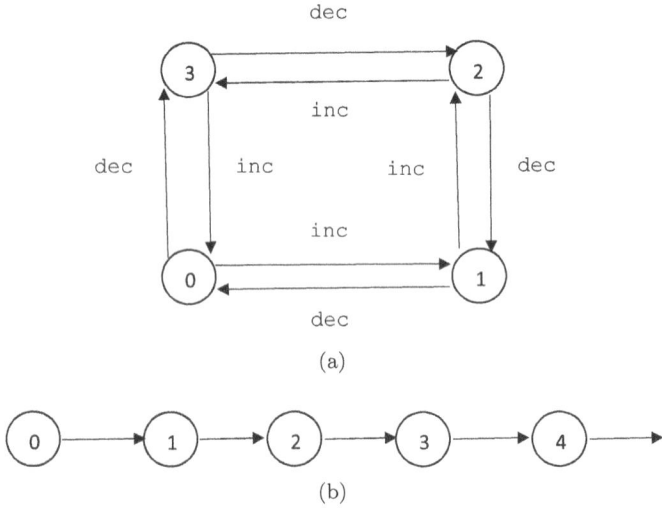

Fig. 2.1.   (a) A transition system for counting modulo 4. (b) Mono-transition system $(\mathbb{N}, \mathrm{succ})$.
Source: Similar to Demri *et al.* (2016, Fig. 3.1).

(3) a binary transition relation $\xrightarrow{a} \subseteq S \times S$ associated with every action name $a \in Act$,

(4) a set PROP of atomic propositions holding in a particular set of states,

(5) a labeling function $L : S \to \mathcal{P}(\mathrm{PROP})$ assigning to every state s the set $L(s)$ of atomic propositions true at s (called the description label of that state),

(6) a designated initial state s called the root of $\mathcal{T}$.

A simple example is a transition system with four states modeling the development of an integer variable modulo 4 (Fig. 2.1(a)). The transition system has two actions, inc for increment and dec for decrement, and no atomic propositions. The initial states are depicted by an incoming arrow. In mono-transition systems, the action label is replaced by a binary relation R on S. For example, in Fig. 2.1(b), the mono-transition system $(\mathbb{N}, \mathrm{succ})$ generates $0, 1, 2, \ldots$ with the successor function succ on $\mathbb{N}$.

A path π in a transition system $\mathcal{T}$ is a finite or an infinite sequence of states and names of actions, which transform every state into its successor, $s_0 \xrightarrow{a_0} s_1 \xrightarrow{a_1} s_2 \cdots$. The path is rooted in s_0. A path with n transitions has length $|\pi| = n$. A finite path in a transition system is a cycle if its first and last states coincide. A loop is a cycle of length 1 with $s \xrightarrow{a} s$.

A transition system is

(1) acyclic if it does not contain cycles,
(2) forest-like if it is acyclic and every state has at most one predecessor state,
(3) tree-like if it is forest-like, in which at most one state has no predecessor states. Such a state is called root and the transition system is called tree.

A computation or trace in a transition system $T = (S, \{\xrightarrow{a}\}_{a \in Act}, L)$ is a sequence of state descriptions and actions along a path with $L(s_0) \xrightarrow{a_0} L(s_1) \xrightarrow{a_1} L(s_2) \cdots$.

2.1. Basic Modal Logic

Basic modal logic (BML) is the simplest temporal logic for reasoning about computational transition systems. Each transition relation of a transition system is associated with a modal operator, stating what must be true in all successors of the current state or what may be true in some successors of the current state: The notation AX_a relates to all paths (A) starting from the current state to the next state (X) with action a. The notation EX_a relates to some path (E) starting from the current state to the next state (X) with action a.[2]

2.1.1. *Syntax of BML*

In the formal language of BML, a BML formula is inductively defined by

$$\varphi := p|\bot|\neg\varphi|\varphi \wedge \varphi| \ EX_a\varphi,$$

---

[2]BML is a variant of the modal logic.

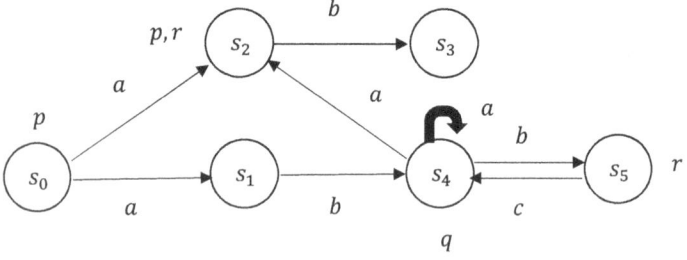

Fig. 2.2. Transition system for proving the truth of certain BML formula. *Source*: After Demri *et al.* (2016, Fig. 5.1).

with $p \in \mathrm{PROP}$ (set of propositions) and $a \in Act$ (set of actions). The other logical connectives are defined as usual. The dual modal operator of EX_a is defined by $\mathrm{AX}_a := \neg \mathrm{EX}_a \neg \varphi$ for all $a \in Act$.

2.1.2.   *Semantics of BML*[3]

The semantics of BML relates to the truth of a formula φ at a state s of an ITS $\mathcal{T} = (S, \left\{ \xrightarrow{a} \right\}_{a \in Act}, L)$, i.e., $\mathcal{T}, s \models \varphi$, which can be defined inductively by

$$\mathcal{T}, s \not\models \bot;$$
$$\mathcal{T}, s \models p \qquad \text{iff } p \in L(s);$$
$$\mathcal{T}, s \models \neg \varphi \quad \text{iff } \mathcal{T}, s \not\models \varphi;$$
$$\mathcal{T}, s \models \varphi \wedge \psi \text{ iff } \mathcal{T}, s \models \varphi \text{ and } \mathcal{T}, s \models \psi;$$
$$\mathcal{T}, s \models \mathrm{EX}_a \varphi \text{ iff } \mathcal{T}, s' \models \varphi \text{ for some } s' \in S \text{ with } s \xrightarrow{a} s'.$$

According to the definition of AX_a, it follows that

$$\mathcal{T}, s \models \mathrm{AX}_a \varphi \text{ iff } \mathcal{T}, s' \models \varphi \text{ for every } s' \in S \text{ with } s \xrightarrow{a} s'.$$

For example, the truth of formula $\mathrm{AX}_a \bot \wedge \mathrm{AX}_b p \wedge \neg \mathrm{EX}_b p$ at state s_3 of a transition system $\mathcal{T}$ can be derived with the graph of $\mathcal{T}$ in Fig. 2.2. The truth of formula $p \wedge \neg q \wedge \mathrm{EX}_a(\neg p \wedge \neg r) \wedge \mathrm{EX}_a(p \wedge r) \wedge \mathrm{AX}_a \neg q \wedge \mathrm{AX}_b \bot$ is derived at state s_0.

---

[3]The semantics of ITSs deliver the possible world semantics of Kripke models for modal logic in computer science. For Kripke models, compare, for example, Hintikka (1962).

```
procedure MC_BML(T,s,φ)

    Case φ of

    p:      return p ∈ L(s)

    ¬φ':    return (not MC_BML(T,s,φ'))

    φ₁ ∧ φ₂:    return ( MC_BML(T,s,φ₁) and MC_BML(T,s,φ₂))

    EXφ':

        if ∃s' ∈ R(s) with MC_BML(T,s,φ') = true then

            return true

        end if

        return false

    end case

end procedure
```

Fig. 2.3. Algorithm MC_{BML} for model checking of BML formulae.
Source: Demri *et al.* (2016, p. 117).

The semantic definitions of satisfiability, validity of a model, logical validity, and logical implication for BML formulae can be introduced with respect to the definition of $\models$. The logical problems of model checking and validity testing cannot always be reduced to each other at low computational costs. The algorithm MC_{BML} (Fig. 2.3) solves the (local) model-checking problem in polynomial time. It is a straightforward implementation of the satisfaction relation $\models$. MC_{BML} works as a procedure for the recursive top-down labeling of the states by subformulae with truth values. It can be proven that, for a given BML formula φ, a finite transition system $\mathcal{T}$, and a state s in $\mathcal{T}$, the algorithm MC_{BML} returns the value true iff $\mathcal{T}, s \models \varphi$. Furthermore, the satisfiability problem of BML is decidable. It can also be proven that any satisfiable BML formula is satisfied at the root of a finite tree-like ITS.

2.1.3. *Axiomatic system* $AxSys_{BML}$ *for BML*

An axiomatic system $AxSys_{BML}$ for BML is given by the axioms of propositional logic (PL) extended by axiom (K) for basic AX and two rules of inference.

Axioms of AxSys$_{\mathbf{BML}}$:

(PL) All axioms of PL

(K) $AX\varphi \wedge AX(\varphi \to \psi) \to AX\psi$

Rules of inference:

(MP) Modus ponens: $\dfrac{\vdash\varphi,\ \vdash\varphi\to\psi}{\vdash\psi}$

(Nec) Necessitation: $\dfrac{\vdash\varphi}{\vdash AX\varphi}$

2.2.   Linear-Time Temporal Logic

Linear temporal logic (LTL) considers single computations in transition systems which are represented by linear models over infinite
sequences of states.[4] Thus, LTL does not only relate to local properties, such as BML. But LTL cannot express branching-time
properties, such as computation tree logic (CTL), which will be studied in the following section. The language of LTL extends the classical
PL with temporal modalities for future states in computations. They
can be illustrated for current states φ in linear sequences of states.

The temporal operator $X\varphi$ states that the next state satisfies φ. For
example, in a computer program, the formula alert $\to X$ halt expresses
that after a current state of alert, the program should halt in the next
state. The formula $F\varphi$ states that φ will be true sometime in the future
or at the current state. $G\varphi$ means that all future states including the
current state satisfy φ. $G\varphi$ and $\neg F\neg\varphi$ are equivalent. The binary
time operator Until ($\bigcup$) in the formula $\varphi\bigcup\psi$ expresses that φ will be
true from now until ψ becomes true at some future state.

2.2.1.   *Syntax of LTL*

In the formal language of LTL, an LTL formula is inductively defined
by

$$\varphi\psi := p|\bot|\neg\varphi|\varphi \wedge \varphi|X\,\varphi|F\varphi|G\varphi|\varphi\bigcup\psi,$$

with $p \in \mathrm{PROP}$ (set of propositions).

---

[4]Gabbay *et al.* (1980); Pnuelli (1977).

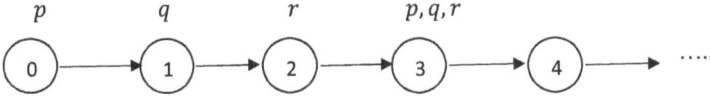

Fig. 2.4. Example of a LTL model.

2.2.2. Semantics of LTL

The models of LTL are infinite computations, which are illustrated as sequences of labels with atomic propositions, as shown in Fig. 2.4.[5]

Formally, a linear model of LTL is defined by $\sigma : \mathbb{N} \to \mathcal{P}(\text{PROP})$. For a linear model σ, a position $i \in \mathbb{N}$ and a formula φ, the satisfaction relation $\models$ can be defined inductively by

$$\sigma, i \not\models;$$
$$\sigma, i \models p \quad \text{iff} \quad p \in \sigma(i);$$
$$\sigma, i \models \neg\varphi \quad \text{iff} \quad \sigma, i \not\models \varphi;$$
$$\sigma, i \models \varphi \wedge \psi \quad \text{iff} \quad \sigma, i \models \varphi \text{ and } \sigma, i \models \psi;$$
$$\sigma, i \models \mathrm{X}\varphi \quad \text{iff} \quad \sigma, i+1 \models \varphi;$$
$$\sigma, i \models \mathrm{F}\varphi \quad \text{iff} \quad \text{there is } j \geq i \text{ such that } \sigma, j \models \varphi;$$
$$\sigma, i \models \mathrm{G}\varphi \quad \text{iff} \quad \text{for all } j \geq i \text{ we have } \sigma, j \models \varphi;$$
$$\sigma, i \models \varphi \bigcup \psi \quad \text{iff} \quad \text{there is } j \geq i \text{ such that } \sigma, j \models \psi \text{ and}$$
$$\sigma, k \models \varphi \text{ for all } k \text{ with } i \leq k < j.$$

Truth $\sigma \models \varphi$ of an LTL formula φ in an LTL model σ is defined by $\sigma, 0 \models \varphi$.

2.2.3. Decidability problems of LTL[6]

Satisfiability and validity are associated with decision problems.

---

[5]Demri *et al.* (2016, p. 155).
[6]Demri *et al.* (2016, p. 159).

Satisfiability problem SAT(LTL) for LTL:

Input: LTL formula φ

Question: Is there any model σ with $\sigma \models \varphi$?

Validity problem VAL(LTL) for LTL:

Input: LTL formula φ

Question: Is the case that $\models \varphi$?

For checking the satisfiability of an LTL formula φ, one can try to find a small satisfiability witness. A brute force algorithm generates all sequences of subsets in the closure $cl_{LTL}(\varphi)$ of φ with a length of at most $2^{|\varphi|} + |\varphi| \cdot 2^{|\varphi|}$ and checks whether one of them is a small satisfiability witness. This kind of algorithm is a decision procedure for LTL satisfiability with double exponential time of computation. Another procedure starts with a guess of a sequence with a length of at most $2^{|\varphi|} + |\varphi| \cdot 2^{|\varphi|}$ for checking whether it is a small satisfiability witness. Figure 2.5 is a nondeterministic algorithm for checking whether a given LTL formula has a small satisfiability witness.

The algorithm in Fig. 2.5 for satisfiability checking of LTL formulae is correct and needs a space which is polynomial in the size of the input formula.

For example, for the LTL formula $\varphi = GFp \wedge GFq$ und $\Gamma :=$ $\{\varphi, GFp, GFq, Fp, Fq\}$, the following sequence can be checked by the algorithm in Fig. 2.5 to be a small satisfiability witness with $i = 2$ and $j = 4$:

$$\Gamma \cup \{\neg p, \neg q\}, \quad \Gamma \cup \{\neg p, \neg q\}, \quad \Gamma \cup \{p, \neg q\}, \quad \Gamma \cup \{\neg p, \neg q\},$$
$$\Gamma \cup \{\neg p, q\}, \quad \Gamma \cup \{p, q\}, \quad \Gamma \cup \{p, \neg q\}.$$

Until now, the temporal operators of LTL only refer to the future. They can be extended by the counterparts of the past. The "next" operator X has an analog in past time, which is called the "previous" operator Y ("yesterday"): $Y\varphi$ states that the previous state satisfies φ. The "until" operator $\bigcup$ has an analog in past time, which is called the "since" operator S: The binary time operator S in the formula $\varphi S\psi$ expresses that ψ is true at some past position, and since then, φ holds true.

guess $i \in \left[0, 2^{|\varphi|}\right]$ and $l \in \left[1, |\varphi| \cdot 2^{|\varphi|}\right]$

guess $\Gamma \subseteq cl_{LTL}(\varphi)$ such that $\varphi \in \Gamma$ and Γ is maximally consistent

$j \leftarrow 0$

while $j < i$ do

    guess maximally consistent Γ' such that (Γ, Γ') is one-step consistent;

    $j \leftarrow j + 1$

    $\Gamma \leftarrow \Gamma'$

end while

$\Gamma_f \leftarrow \Gamma$

$j \leftarrow 0$

$\Delta_U \leftarrow \emptyset$

$\Delta_U^+ \leftarrow \emptyset$

while $j < l$ do

    $\Delta_U \leftarrow \Delta_U \cup \{\psi_1 U \psi_2 \in cl_{LTL}(\varphi) | \psi_1 U \psi_2 \in \Gamma\}$

    $\Delta_U' \leftarrow \Delta_U' \cup \{\psi_1 U \psi_2 \in cl_{LTL}(\varphi) | \psi_2 \in \Gamma\}$

    guess maximally consistent Γ' such that (Γ, Γ')

    $j \leftarrow j + 1$

    $\Gamma \leftarrow \Gamma'$

end while

if $\Gamma = \Gamma_f$ and $\Delta_U \subseteq \Delta_U'$ then

    accept

else

    abort

end if

Fig. 2.5.   Algorithm for guessing a small satisfiability witness.
Source: Demri *et al.* (2016, p. 170).

2.2.4.   *Axiomatic system AxSys$_{\text{LTL}}$ for LTL*[7]

An axiomatic system AxSys$_{\text{LTL}}$ for LTL is given by the axioms of PL extended by axioms for the temporal operators X, G, and $\bigcup$ and three rules of inference, as follows.

Axioms of AxSys$_{\text{LTL}}$:

(PL)   All axioms of PL
(K$_{\text{X}}$) $\text{X}(\varphi \to \psi) \to (\text{X}\varphi \to \text{X}\psi)$
(SER) $\text{X}\top$
(FUNC) $\text{X}\neg\varphi \leftrightarrow \neg\text{X}\varphi$
(PostFP$_{\text{G}}$) $\text{G}\varphi \to (\varphi \wedge \text{XG}\varphi)$
(GFP$_{\text{G}}$) $\text{G}(\psi \to (\varphi \wedge \text{X}\psi)) \to (\psi \to G\varphi)$
(PreFP$_{\bigcup}$) $(\psi \vee (\varphi \wedge \text{X}(\varphi \bigcup \psi))) \to (\varphi \bigcup \psi)$
(LFP$_{\bigcup}$) $\text{G}((\psi \vee (\varphi \wedge \text{X}\chi) \to \chi) \to ((\varphi \bigcup \psi) \to \chi)$

Rules of Inference in AxSys$_{\text{LTL}}$:

(MP)   Modus ponens: $\dfrac{\vdash\varphi, \ \vdash\varphi\to\psi}{\vdash\psi}$
(Nec$_{\text{X}}$)   Necessitation for X: $\dfrac{\vdash\varphi}{\vdash\text{X}\varphi}$
(Nec$_{\text{G}}$)   Necessitation for G: $\dfrac{\vdash\varphi}{\vdash\text{G}\varphi}$

The axiomatic system AxSys$_{\text{LTL}}$ is sound and complete because for every finite set of LTL formulae $\{\varphi_1, \ldots, \varphi_n, \varphi\}$, it holds that

$$\varphi_1, \ldots, \varphi_n \vdash_{\text{AxSys}_{\text{LTL}}} \psi \quad \text{if and only if} \quad \varphi_1, \ldots, \varphi_n \models \psi.$$

Therefore, every set of LTL formulae is satisfiable if and only if it is consistent in AxSys$_{\text{LTL}}$. Especially, $\vdash_{\text{AxSys}_{\text{LTL}}} \psi$ if and only if ψ is a valid LTL formula ($\models \psi$).

2.3.   Computation Tree Logic

LTL is restricted to linear-time models with single computations of transition systems. In CTL, the potential of temporal operators is

---

[7]LTL is a standard formalism to specify the behavior of computer systems for formal verification. Cf. Demri *et al.* (2016, Section 6.8).

combined with the ability to quantify over paths of computation, which leads to branching-time logic.[8] Examples of combinations of temporal operators with path quantifiers have already been introduced. The existence (E) of a run such that the next position (X) satisfies p can be written as $E\,X\,p$ and corresponds to modal operator EX in the BML. The statement that for all (A) runs, the next position (X) satisfies p can be written as $A\,X\,p$ and corresponds to the modal operator AX. The existence (E) of a run such that a future position (F) satisfies p can be written as EFp. It means that there is a reachable state which satisfies p. In the formula $E(p\bigcup q)$, at first, a run is existentially quantified (E) to witness the satisfaction of $p\bigcup q$. The statement that for all runs (A) and for all positions, p holds true can be written as AGp. Thus, the combination AG can be understood as a temporal operator in branching-time logic consisting of a universal quantifier over paths and a universal quantifier over positions.

2.3.1. *Temporal logic of reachability*

In a transition system $T = (S, \{\overset{a}{\longrightarrow}\}_{a\in Act}, L)$, the reflexive and transitive closure $s\overset{a}{\longrightarrow}{}^{*}t$ of the transition relation $\overset{a}{\longrightarrow}$ means that there exists a finite path $s_0 \overset{a}{\longrightarrow} s_1 \overset{a}{\longrightarrow} \cdots \overset{a}{\longrightarrow} s_n$ with $n \geq 0$, $s = s_0$, and $t = s_n$. Thus, $\overset{a}{\longrightarrow}{}^{*}$ denotes the reachability relation for transitions of type a. The reachability relation can be associated with a reachability operator, resp. reachability modality, EF_a, which means forward reachability in the future (F) along some computation with a.

The temporal logic of reachability (TLR) is an extension of BML with the reachability operators EF_a for all transition relations $\overset{a}{\longrightarrow}$ of transition system T.[9] The syntax of TLR is inductively defined by

$$\varphi := p|\bot|\neg\varphi|(\varphi \wedge \varphi)|EX_a\varphi|EF_a\varphi \quad \text{with } p \in \text{PROP}.$$

As in BML, the dual of each modal operator can be defined by

$$AX_a\varphi := \neg EX_a\neg\varphi,$$
$$AG_a\varphi := \neg EF_a\neg\varphi.$$

---

[8]Clarke and Emerson (1981), Emerson (1990).
[9]Ben-Ari *et al.* (1981).

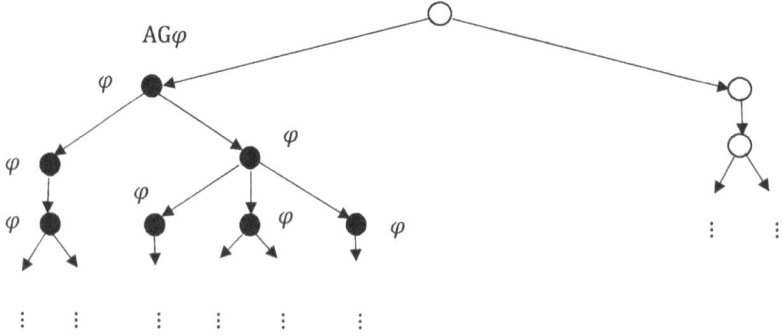

Fig. 2.6. φ is true at every reachable state (AG_φ).
Source: Demri *et al.* (2016, Fig. 7.2).

The formulae of TLR are interpreted over rooted transition systems. The satisfaction relation $\models$ is defined by the conditions of BML extended by the case

$$\mathcal{T}, s \models \mathrm{EF}_a\varphi \quad \text{if and only if } \mathcal{T}, s \models \varphi \text{ for some } r \in S \text{ with } s \xrightarrow{a}{}^* r.$$

Intuitively, $\mathrm{EF}_a\varphi$ is true at the state s if φ is true at some state r reachable from s along the transition relation $\xrightarrow{a}$.
According to the definition AG_a, it follows that

$$\mathcal{T}, s \models \mathrm{AG}_a\varphi \quad \text{if and only if } \mathcal{T}, s \models \varphi \text{ for all } r \in S \text{ with } s \xrightarrow{a}{}^* r.$$

Figure 2.6 illustrates that $\mathrm{AG}_a\varphi$ is true at the state s if φ is true at every state r reachable from s along the transition relation $\xrightarrow{a}$ of a transition system.

In BML, the satisfiability of TLR is decidable. An axiomatic system $\mathrm{AxSys_{TLR}}$ for the set of valid formulae of TLR extends $\mathrm{AxSys_{BML}}$ with an axiom to guarantee the seriality of the transition relation and axioms for the operator AG.[10]

---

[10]Demri *et al.* (2016, p. 253).

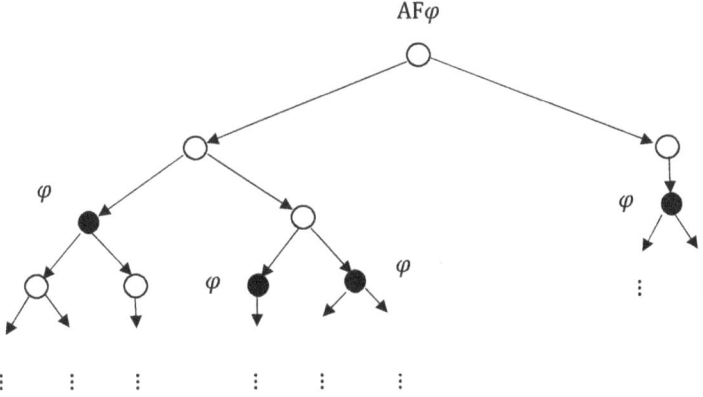

Fig. 2.7. Illustration of $AF\varphi$.
Source: Demri *et al.* (2016, Fig. 7.3).

Axioms of $AxSys_{TLR}$:

(BML) All axioms of BML logic $AxSys_{BML}$
(K_{AX}) $AX(\varphi \to \psi) \to (AX\varphi \to AX\psi)$
(SER) $EX\top$
($PostFP_{AG}$) $AG\varphi \to (\varphi \land AXAG\varphi)$
(GFP_{AG}) $AG(\psi \to (\varphi \land AX\psi)) \to (\psi \to AG\varphi)$

Rules of Inference in $AxSys_{LTL}$:

(MP) Modus ponens: $\dfrac{\vdash\varphi,\ \vdash\varphi\to\psi}{\vdash\psi}$

(Nec_X) Necessitation for AX: $\dfrac{\vdash\varphi}{\vdash AX\varphi}$

(Nec_G) Necessitation for AG: $\dfrac{\vdash\varphi}{\vdash AG\varphi}$

According to $PostFP_{AG}$, $AG\varphi$ is a post-fixpoint of the operator Γ_{AG} with $\Gamma_{AG}(\vartheta) = \varphi\land AX\vartheta$. GFP_{AG} means that $AG\varphi$ is the greatest post-fixpoint of the operator Γ_{AG}.

2.3.2. *Computation tree logic*

TLR only claims reachability on some computations. CTL is an extension of TLR which refers to all computations starting from the current state, which is considered by the operator AF. Figure 2.7 illustrates $AF\varphi$ with an example.

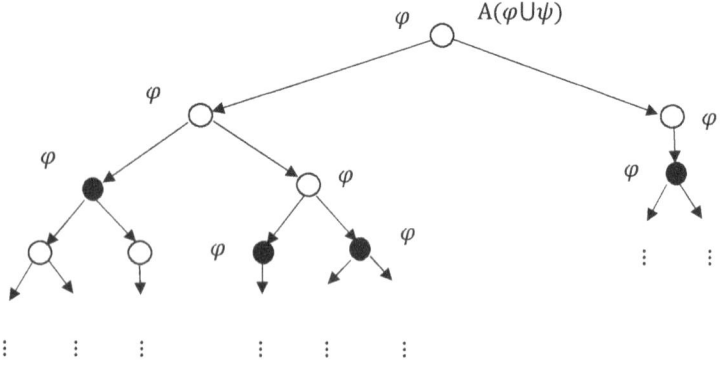

Fig. 2.8.   Illustration of A($\varphi \bigcup \psi$).
Source: Demri *et al.* (2016, Fig. 7.3).

The CTL is an extension of BML with modal operators for existential and universal reachability. They combine the existential, resp. universal, path quantifier with the Until operator. The syntax of CTL is inductively defined by

$$\varphi := p|\bot|\neg\varphi|(\varphi \wedge \varphi)|\text{EX}\varphi\left|\text{E}\left(\varphi \bigcup \varphi\right)\right|\text{A}\left(\varphi \bigcup \varphi\right) \quad \text{with } p \in \text{PROP}.$$

In A($\varphi \bigcup \psi$), the Until property ($\bigcup$) is combined with universal path quantification (A). In this case, ψ must eventually hold for all paths, and φ holds at every moment before that on all these paths (Fig. 2.8).

In CTL, the existential reachability operator of TLR can be defined as EF$\varphi := \text{E}(\top \bigcup \varphi)$. The universal reachability operator is AF$\varphi := \text{A}(\top \bigcup \varphi)$. The dual EG$\varphi := \neg\text{AF}\neg\varphi$ states the existence of a computation in which φ holds at every moment. Furthermore, AG$\varphi := \neg\text{EF}\neg\varphi$.

The satisfaction relation $\models$ is defined by the conditions of TLR extended by the following cases:

$\mathcal{T}, s \models \text{E}(\varphi_1 \bigcup \varphi_2)$ iff there is a path π which starts at s and $i \geq 0$ such that $\pi(0) = s$, $\mathcal{T}, \pi(i) \models \varphi_2$, and for every $j \in [0, i-1]$ it is $\mathcal{T}, \pi(j) \models \varphi_1$.

$\mathcal{T}, s \models \text{A}(\varphi_1 \bigcup \varphi_2)$ iff for all paths π such that $\pi(0) = s$, there is $i \geq 0$ such that $\mathcal{T}, \pi(i) \models \varphi_2$, and for every $j \in [0, i-1]$, it is $\mathcal{T}, \pi(j) \models \varphi_1$.

The set of valid formulae of CTL is axiomatized by AxSys$_{\text{CTL}}$.[11]

(BML) All axioms of BML logic AxSys$_{\text{BML}}$
(K$_{\text{AX}}$) AX($\varphi \to \psi$) $\to$ (AX$\varphi \to$ AXψ)
(SER) EX$\top$
(PreFP$_{\text{E}\bigcup}$) ($\psi \vee (\varphi \wedge \text{EXE}(\varphi \bigcup \psi)))$ $\to$ E($\varphi \bigcup \psi$)
(PreFP$_{\text{A}\bigcup}$) ($\psi \vee (\varphi \wedge \text{AXA}(\varphi \bigcup \psi)))$ $\to$ A($\varphi \bigcup \psi$)
(LFP$_{\text{E}\bigcup}$) AG(($\psi \vee (\varphi \wedge \text{EX}\chi) \to \chi$) $\to$ (E($\varphi \bigcup \psi$) $\to \chi$)
(LFP$_{\bigcup}$) AG(($\psi \vee (\varphi \wedge \text{AX}\chi) \to \chi$) $\to$ (A($\varphi \bigcup \psi$) $\to \chi$)

2.4. Full Computation Tree Logic

In TRL and CTL, certain temporal operators, such as X and $\bigcup$, were already combined with path quantifiers. In full computation tree logic (CTL*), the definitions of LTL and CTL include state formulae related to states and path formulae related to computations of an ITS. In this sense, CTL* comprises the full branching logic.[12]

2.4.1. *Syntax of CTL**

In the syntax of CTL*, state formulae φ and path formulae ϑ are defined in mutual recursion by

$$\varphi := \bot |p|\neg\varphi|\varphi \wedge \varphi|\text{A}\vartheta \quad \text{with } p \in \text{PROP},$$

$$\vartheta := \varphi|\neg\vartheta|\vartheta \wedge \vartheta|\text{X}\vartheta|\vartheta \bigcup \vartheta.$$

The existential path quantifier is defined by E$\varphi := \neg\text{A}\neg\varphi$.

2.4.2. *Semantics of CTL**

The semantics of CTL* relates to a transition system $\mathcal{T}$ with a satisfaction relation $\models_s$ of state formulae and a satisfaction relation $\models_p$

---

[11]Demri *et al.* (2016, p. 255).

[12]Emerson and Halpern (1983); Demri *et al.* (2016, pp. 2017–2018, 258).

of path formulae:

$\mathcal{T}, s_s \not\models_s \bot$;

$\mathcal{T}, s \models_s p$ iff $p \in L(s)$;

$\mathcal{T}, s \models_s \neg\varphi$ iff $\mathcal{T}, s_s \not\models_s \varphi$;

$\mathcal{T}, s \models_s \varphi \wedge \psi$ iff $\mathcal{T}, s \models_s \varphi$ and $\mathcal{T}, s \models_s \psi$;

$\mathcal{T}, s \models_s \mathrm{E}\vartheta$ iff there is a path π starting at s

                 such that $\mathcal{T}, \pi \models_s \vartheta$;

$\mathcal{T}, s \models_s \mathrm{A}\vartheta$ iff for all paths π starting at s,

                 it holds that $\mathcal{T}, \pi \models_s \vartheta$;

$\mathcal{T}, \pi \models_p \varphi$ iff $\mathcal{T}, \pi(0) \models_s \varphi$ for state formulae φ;

$\mathcal{T}, \pi \models_p \neg\vartheta$ iff $\mathcal{T}, \pi_s \not\models_s \vartheta$;

$\mathcal{T}, \pi \models_p \vartheta \wedge \vartheta'$ iff $\mathcal{T}, \pi \models_p \vartheta$ and $\mathcal{T}, \pi \models_p \vartheta'$;

$\mathcal{T}, \pi \models_p \mathrm{X}\vartheta$ iff $\mathcal{T}, \pi[1, +\infty] \models_p \vartheta$;

$\mathcal{T}, \pi \models_p \vartheta \bigcup \vartheta'$ iff there is $i \geq 0$ such that $\mathcal{T}, \pi[i, +\infty] \models_p \vartheta'$ and

                 for every $j \in [0, i - 1]$, it is $\mathcal{T}, \pi[j, +\infty] \models_p \vartheta$

Since every state formula of CTL* is also a path formula, it is usual to merge the two sorts and assume that all CTL* formulae are path formulae.

The semantics of CTL* can be restricted to tree-like transition systems. A uniformly branching tree has an equal number of branches at every node. For a given number k, a k-branching tree is a uniformly branching tree with exactly k branches at every node. It can be proven that every satisfiable state formula φ (in negation normal form) of CTL* is satisfiable in a k-branching tree, with $k \leq m + 1$, with m number of existential path quantifiers occurring in φ. Whereas model checking for CTL and TLR can be done in polynomial time, we need exponential time for CTL*. A practical application of model checking in CTL and CTL* is program verification in computer science. Program verification is important to prevent critical states and to guarantee safety.

2.4.3. *Axiomatic system* $\mathrm{AxSys}_{\mathrm{CTL}*}$ *for CTL**

The axioms of $\mathrm{AxSys}_{\mathrm{CTL}*}$ for the full branching logic CTL* includes the axioms of $\mathrm{AxSys}_{\mathrm{LTL}}$ for path formulae, axioms for the path quantifiers, and an axiom for the interaction of A and X.

Axioms of AxSys$_{\text{CTL}*}$:

All axioms of AxSys$_{\text{LTL}}$

$$A(\varphi \to \psi) \to (A\varphi \to A\psi)$$

$$A\varphi \to AA\varphi$$

$$A\varphi \to \varphi$$

$$A\varphi \to AE\varphi$$

$p \to Ap$, for each atomic proposition p

$$AX\varphi \to XA\varphi$$

Rules of Inference in AxSys$_{\text{LTL}}$:

(MP) Modus ponens: $\frac{\vdash\varphi,\ \vdash\varphi\to\psi}{\vdash\psi}$

(Nec$_{\text{AX}}$) Necessitation for AX: $\frac{\vdash\varphi}{\vdash AX\varphi}$

(Nec$_{\text{AG}}$) Necessitation for AG: $\frac{\vdash\varphi}{\vdash AG\varphi}$

(Nec$_{\text{A}}$) Necessitation for A: $\frac{\vdash\varphi}{\vdash A\varphi}$

The axiomatic system AxSys$_{\text{CTL}*}$ is sound and complete with respect to the class of all (generalized) branching structures.[13]

2.4.4. Ockhamist CTL*

LTL was extended by past-time operators which do not change the class of models. In a branching time, several aspects must be considered for representing the past. With respect to Ockham's philosophy of time (compare Sections 1.1 and 1.2), a tree-like representation with a finite and linear past, but infinite and possibly branching ("open"), future is assumed. Figure 2.9 is an illustration of the Ockhamist philosophy of time in the framework of extended CTL*.[14]

---

[13]Reynolds (2001).
[14]Prior (1967) and Demri *et al.* (2016, p. 250).

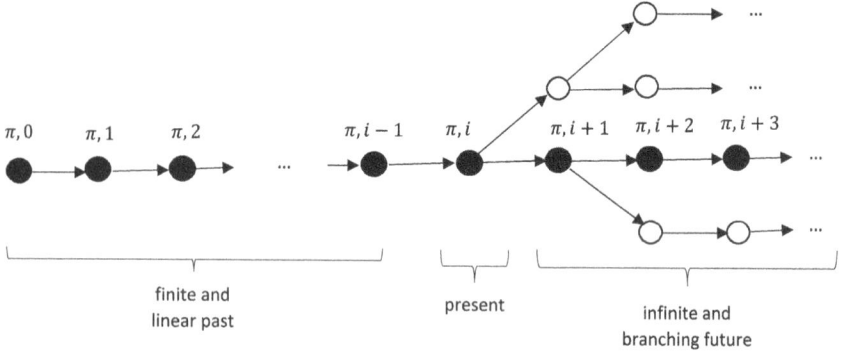

Fig. 2.9.   A path in Ockhamist CTL*.

The extension of CTL* with this kind of Ockhamist past needs an extended syntax with past operators Y and S:

$$\varphi := p|\bot|\neg\varphi| \left(\varphi \wedge \varphi\right) |\mathrm{X}\varphi| \left(\varphi \bigcup \psi\right) |\mathrm{E}\varphi|\mathrm{A}\varphi|\mathrm{Y}\varphi|(\varphi \mathrm{S}\psi)$$

with $p \in \mathrm{PROP}$.

In the semantics of the past-extended CTL*, the truth of formulae is evaluated for a path and a position along the path:

$\mathcal{T}, \pi, i \models p$ iff $p \in L(\pi(i))$;

$\mathcal{T}, \pi, i \models \neg\varphi$ iff $\mathcal{T}, \pi, i \not\models \varphi$;

$\mathcal{T}, \pi, i \models \varphi \wedge \psi$ iff $\mathcal{T}, \pi, i \models \varphi$ and $\mathcal{T}, \pi, i \models \psi$;

$\mathcal{T}, \pi, i \models \mathrm{E}\varphi$ iff there is a path $\mathcal{T}, \pi'$ such that
         $\mathcal{T}, \pi[0, i] = \pi'[0, i]$ and $\mathcal{T}, \pi', i \models \varphi$;

$\mathcal{T}, \pi, i \models \mathrm{X}\varphi$ iff $\mathcal{T}, \pi, i + 1 \models \varphi$;

$\mathcal{T}, \pi, i \models \varphi \bigcup \psi$ iff there is $j \geq i$ such that $\mathcal{T}, \pi, j \models \psi$, and
         for every $k \in [i, j - 1]$, it is $\mathcal{T}, \pi, k \models \varphi$;

$\mathcal{T}, \pi, i \models \mathrm{Y}\varphi$ iff $i > 0$ and $\mathcal{T}, \pi, i - 1 \models \varphi$;

$\mathcal{T}, \pi, i \models \varphi \mathrm{S}\psi$ iff there is $j \leq i$ such that $\mathcal{T}, \pi, j \models \psi$, and
         for every $k \in [j - 1, i]$, it is $\mathcal{T}, \pi, k \models \varphi$.

Chapter 3

Proof-Theoretical Foundations of Temporal Logic

In the age of digitalization and Big Data, security and trust in information flow and computational systems have become urgent demands of technology and societal acceptance. Therefore, specification and verification, model-checking, and satisfiability testing of hardware and software are great challenges in computer science. Temporal logic plays a substantial role in model-checking and verification technology. But users in the digital world are often less aware that verification, model-checking, and satisfiability are deeply rooted in the logical foundations of proof theory. In this chapter, we continue with early studies on the philosophical and proof-theoretical foundations of computer science with respect to temporal logic.[1]

The first proof systems for temporal logic were developed with Hilbert-style axiomatic calculi. In this case, proof systems are constituted with several axioms and rules of inference (compare Chapter 2). But axiomatic systems of this style are only useful for formalization of what has already been proved. They are not appropriate for the search of proofs which are demanded for verification problems in computer science. For proof searching, we need guidance by the propositions and theorems to be proved, which is missing in Hilbert-style systems.

---

[1]Mainzer *et al.* (2018, 2022).

Here, Gentzen-type systems come in. In the beginning of the 1930s, Gerhard Gentzen introduced a sequent calculus which overcomes the lack of guidance of Hilbert-style systems by a special notation. At any step of derivation, the antecedent of a sequent shows the open assumptions on which the formulas in the consequent depend. If the proof search is terminating, then the decision problems in formal systems can be solved. In the following, we start the proof analysis in temporal logic with the Gentzen sequent calculus, which is the historical "mother" of several related proof-theoretical systems, such as tableaux-based, automata-based, and game-based calculi. They have different advantages for the formal verification of efficient algorithmic methods solving verification problems. In these cases, the decision problems for temporal logic can be reduced to decision problems about tableaux, automata, and games, respectively. For example, typical challenges are problems checking whether an automaton accepts a certain computation or whether it recognizes a certain language.

3.1.   Gentzen Calculus and Temporal Logic

The proof-theoretic treatment of temporal logics (indeed, of many other families of nonclassical logics as well) within Gentzen's original framework that was tailored for classical logic (and adaptable to intuitionistic logic) is notoriously difficult. Since the 1980s, a lot of hard work and ingenuity has been and still is being devoted to capture these logics with adequate, well-behaved "sequent-style" calculi.

As far as temporal logic systems are concerned, difficulties are more challenging than those encountered in plain modal logic systems already at the level of the minimal system TL: This is due to the intertwining of the two □-like operators H and G (and the two associated ◇-like operators P and F) of TL, as is evident from the two characteristic tautologies $\varphi \to \text{GP}\varphi$ and $\varphi \to \text{HF}\varphi$ of this logic.

In the following, after two preliminary sections in which we explain what a sequent-style calculus is and which properties are requested to make a sequent-style calculus into a "good" one, we illustrate a selected number of interesting contributions concerning the proof-theoretic, Gentzen-type approach to TL, LTL, and CTL and discuss some problems as well.

3.1.1. Gentzen's calculus for classical propositional logic in a nutshell

To answer the question "What is a sequent-style (or Gentzen-type) calculus?" it is convenient to start by recalling the basics of Gentzen's original sequent calculus LK for classical predicate logic, introduced in 1935.[2] For the sake of simplicity, we confine to LK's propositional fragment LK_p (primitive logical operators: $\neg, \vee, \wedge$, and $\rightarrow$) since the temporal logics we are considering here are extensions of classical propositional logic and do not feature individual quantification.

A sequent, according to Gentzen's original definition, is a syntactic object of the form

$$\Gamma \Rightarrow \Delta,$$

where Γ (the antecedent of the sequent) and Δ (the consequent) are two finite, possibly empty lists, say $\varphi_1, \ldots \varphi_n$ and $\psi_1, \ldots \psi_m$ ($n, m \geq 0$), of not necessarily distinct formulae.

According to Gentzen, the intended meaning of a sequent $\Gamma \Rightarrow \Delta$ is represented by the conditional formula $\bigwedge \Gamma \rightarrow \bigvee \Delta$, whose antecedent is the conjunction of the formulae in Γ (an arbitrarily fixed tautology $\top$ if Γ is empty) and whose consequent is the disjunction of the formulae in Δ (an arbitrarily fixed contradiction $\bot$ if Δ is empty). Thus, from a classical-semantics point of view, a sequent is true if, whenever all formulae in its antecedent are true, at least one formula in its consequent is also true.

Note that the list-forming comma "," acts therefore as a "structural" logical operator having a "positional meaning," namely a conjunctive one in the antecedent and a disjunctive one in the consequent.

The calculus LK_p features axioms (the so-called "initial sequents")

$$\varphi \Rightarrow \varphi$$

and (one- or two-premise) inference rules. The latter are divided into "structural" and "logical" rules.

---

[2] Gentzen (1935, pp. 176–210).

The structural rules, Exchange (E), Weakening (W), Contraction (C), and the "cut rule" (cut), do not deal with any specific logical operator. E, W, C come in pairs and produce structural modifications in the antecedent or in the consequent of a sequent; the cut rule represents instead a generalized form of the *modus ponens* or of the transitivity rule for the conditional:

$$\frac{\Gamma \Rightarrow \Delta}{\varphi, \Gamma \Rightarrow \Delta} \; W_l, \qquad\qquad \frac{\Gamma \Rightarrow \Delta}{\Gamma \Rightarrow \Delta, \varphi} \; W_r,$$

$$\frac{\Gamma_1, \varphi, \psi, \Gamma_2 \Rightarrow \Delta}{\Gamma_1, \psi, \varphi, \Gamma_2 \Rightarrow \Delta} \; E_l, \qquad\qquad \frac{\Gamma \Rightarrow \Delta_1, \varphi, \psi, \Delta_2}{\Gamma \Rightarrow \Delta_1, \psi, \varphi, \Delta_2} \; E_r,$$

$$\frac{\varphi, \varphi, \Gamma \Rightarrow \Delta}{\varphi, \Gamma \Rightarrow \Delta} \; C_l, \qquad\qquad \frac{\Gamma \Rightarrow \Delta, \varphi, \varphi}{\Gamma \Rightarrow \Delta, \varphi} \; C_r,$$

$$\frac{\Gamma \Rightarrow \Delta, \varphi \quad\quad \varphi, \Pi \Rightarrow \Sigma}{\Gamma, \Pi \Rightarrow \Delta, \Sigma} \; \text{cut}.$$

The logical rules, in turn, come in pairs for each logical operator $*$ (here, one of $\neg, \vee, \wedge, \to$) and introduce in the antecedent ($*_l$), resp. in the consequent ($*_r$), of the conclusion, a complex formula whose main operator is $*$:

$$\frac{\Gamma \Rightarrow \Delta, \varphi}{\neg\varphi, \Gamma \Rightarrow \Delta} \; \neg_l, \qquad\qquad \frac{\varphi, \Gamma \Rightarrow \Delta}{\Gamma \Rightarrow \Delta, \neg\varphi} \; \neg_r,$$

$$\frac{\varphi_i, \Gamma \Rightarrow \Delta}{\varphi_1 \wedge \varphi_2, \Gamma \Rightarrow \Delta} \; \wedge_l (i=1,2), \qquad\qquad \frac{\Gamma \Rightarrow \Delta, \varphi \quad\quad \Gamma \Rightarrow \Delta, \psi}{\Gamma \Rightarrow \Delta, \varphi \wedge \psi} \; \wedge_r,$$

$$\frac{\varphi, \Gamma \Rightarrow \Delta \quad\quad \psi, \Gamma \Rightarrow \Delta}{\varphi \vee \psi, \Gamma \Rightarrow \Delta} \; \vee_l, \qquad\qquad \frac{\Gamma \Rightarrow \Delta, \varphi_i}{\Gamma \Rightarrow \Delta, \varphi_1 \vee \varphi_2} \; \vee_r (i=1,2),$$

$$\frac{\Gamma \Rightarrow \Delta, \varphi \quad\quad \psi, \Pi \Rightarrow \Sigma}{\varphi \to \psi, \Gamma, \Pi \Rightarrow \Delta, \Sigma} \; \to_l, \qquad\qquad \cdot\frac{\varphi, \Gamma \Rightarrow \Delta, \psi}{\Gamma \Rightarrow \Delta, \varphi \to \psi} \; \to_r.$$

The formula which is displayed in the conclusion of a rule is called the "principal formula," while its subformula (or subformulae) which is displayed in the premise(s) is called the "secondary formula (or formulae)." The lists $\Gamma, \Delta, \ldots$ contain the "context formulae."

It is crucial to note that all the structural and logical rules, with the sole exception of the cut rule, are such that any formula occurring in the premise(s) also occurs as a (proper or improper) subformula of a formula in the conclusion — intuitively, no "piece of information" (formula) is lost in the top-to-bottom direction. On the contrary, the cut rule is not, in such a sense, "analytic": the displayed formula φ occurring in both premises, the so called "cut formula," may not appear in the conclusion and so is generally lost in the top-to-bottom direction.

A formal derivation $\mathcal{D}$ in LK_p of a sequent $\Gamma \Rightarrow \Delta$ is a finite tree of sequents which is locally correct with respect to the axioms (so each leaf node is an axiom) and the inference rules and has the sequent $\Gamma \Rightarrow \Delta$ as "end-sequent," i.e., at the root.

A "cut-free derivation," i.e., one in which there are no applications of the cut rule, thus satisfies the fundamental

> "Subformula property": Any formula occurring somewhere in the derivation is a subformula of a formula occurring in the end-sequent of the derivation.

LK_p turns out to be valid and complete for classical propositional logic. Indeed, given an axiomatic, Hilbert-style calculus C, which is known to be classically valid and complete, it is an easy exercise to prove that if a sequent $\Gamma \Rightarrow \Delta$ is LK_p-derivable, then its associated conditional formula $\bigwedge \Gamma \to \bigvee \Delta$ is C-provable and, conversely, that if a formula φ is C-provable, then the sequent $\Rightarrow \varphi$ is LK_p-derivable.

Gentzen's fundamental result (for the whole LK, here for LK_p), known as the "Hauptsatz" or "cut-elimination theorem," says that the cut rule is redundant; more precisely, Gentzen's purely syntactic proof of this result shows that any LK_p-derivation $\mathcal{D}$ can be effectively transformed (through a finite number of rewriting steps) into a cut-free derivation $\mathcal{D}^*$ of the same end-sequent of $\mathcal{D}$. The calculus is thus fully analytic and allows systematic root-first, bottom-up proof-searching.

There are many provably equivalent variants of Gentzen's original LK_p. In the "multisets variant," the antecedent and the consequent of a sequent are taken to be finite multisets, and not lists, of formulae to the effect that the structural exchange rules can be trivially dispensed with. One may combine this variant with the one in which all the sequents of the form $\varphi, \Gamma \Rightarrow \Delta, \varphi$ (generalized initial sequents) are taken as initial sequents to the effect that (also) the structural weakening rules can be dropped while still being provably admissible.[3]

The most interesting among the multisets variants (which we need in the following, when considering the labeled calculi for TL) is the sequent calculus $G3c_p$, inspired by the contributions of Ketonen and Dragalin,[4] in which there are no structural rules except the cut rule. $G3c_p$ results from the multiset variant of LK_p by modifying it as follows:

- the structural rules Weakening and Contraction are dropped (and Exchange, of course), so only the cut rule is retained;
- the axioms are taken in the generalized form $\varphi, \Gamma \Rightarrow \Delta, \varphi$, with the additional restriction that φ is an atomic formula (including $\top, \bot$);
- the two-premises logical rule $\to_l$, which in LK_p is formulated in the so-called "multiplicative" (context-free) version, is replaced by the corresponding "additive" (context-sharing) version (such as $\wedge_r, \vee_l$ in LK_p):

$$\frac{\Gamma \Rightarrow \Delta, \varphi \qquad \psi, \Gamma \Rightarrow \Delta}{\varphi \to \psi, \Gamma \Rightarrow \Delta} \to'_l;$$

- the two twin rules $\wedge_l (i = 1, 2)$ as well as the two twin rules $\vee_l (i = 1, 2)$ are replaced by the two corresponding multiplicative rules (such as $\to_r$ in LK_p):

$$\frac{\varphi, \psi, \Gamma \Rightarrow \Delta}{\varphi \wedge \psi, \Gamma \Rightarrow \Delta} \wedge'_l, \qquad \frac{\Gamma \Rightarrow \Delta, \varphi, \psi}{\Gamma \Rightarrow \Delta, \varphi \vee \psi} \vee'_r.$$

One can easily prove the equivalence of LK_p and $G3c_p$. Most importantly, one can (less easily) prove that the rules of weakening and contraction are cut-free admissible and that all the logical rules are invertible (with the preservation of the height of derivations and without making use of cut) and, finally, that also $G3c_p$ admits syntactic cut elimination. The additional key property of $G3c_p$, i.e., the

---

[3]Some authors also use to define sequents as being of the form $\Gamma \Rightarrow \Delta$ with Γ, Δ finite, possibly empty sets of formulae. In this way, the contraction rule is trivialized, being hidden in the syntactic notion of a sequent. There are, however, some drawbacks with this variant, which suggest not to adopt it.

[4]Ketonen (1944), Dragalin (1979). See also Troelstra and Schwichtenberg (2000).

absence of the contraction rule (contraction, so to speak, is hidden in the logical rules by the clever choice and mixture of additive and multiplicative formulations), represents, from the point of view of the efficiency of proof-searching, the great advantage of G3c$_p$ over LK$_p$.

One last variant, which is worth recalling here (also for future use), is often referred to as the "Tait-style" variant.[5] It features "one-sided sequents," which are just finite, possibly empty multisets Γ of formulae, to be intuitively interpreted by the corresponding disjunctive formula $\bigvee \Gamma$. Thus, the standard Gentzen sequent $\Gamma \Rightarrow \Delta$ corresponds to the one-sided sequent

$$\neg\Gamma, \Delta$$

(where $\neg\Gamma := \{\neg\varphi \mid \varphi \in \Gamma\}$).

This variant fits well only with classical logic and extensions thereof and benefits from the adoption of a propositional language featuring literals (countably many positive and negative propositional variables $p_0, p_1, \ldots$, resp. $\bar{p}_0, \bar{p}_1, \ldots$) and, as primitive boolean connectives, only the conjunction and the disjunction. The negation $\neg\varphi$ of a formula φ is inductively defined via the double negation and De Morgan laws ($\neg p_i := \bar{p}_i, \neg\bar{p}_i := p_i, \neg(\varphi \vee \psi) := \neg\varphi \wedge \neg\psi, \neg(\varphi \wedge \psi) := \neg\varphi \vee \neg\psi$), while the remaining connectives are defined as usual. The advantage is a substantial economy in the formulation of the calculus. For instance, the Tait version of LK$_p$ is defined as follows:

Axioms (initial sequents): $\quad p, \bar{p}$.

Structural rules:

$$\frac{\Gamma}{\Gamma, \Delta}\ \text{W}, \qquad \frac{\Gamma, \varphi, \varphi}{\Gamma, \varphi}\ \text{C}, \qquad \frac{\Gamma, \varphi \qquad \neg\varphi, \Gamma}{\Gamma}\ \text{cut}.$$

Logical rules:

$$\frac{\Gamma, \varphi \qquad \Gamma, \psi}{\Gamma, \varphi \wedge \psi}\ \wedge, \qquad \frac{\Gamma, \varphi, \psi}{\Gamma, \varphi \vee \psi}\ \vee.$$

---

[5]Tait (1968). Note that William Tait uses sets of formulae instead of multisets of formulae.

3.1.2.   *What is a (good) sequent-style calculus?*

With a few exceptions (notably, intuitionistic logic[6]), most nonclassical logics and families thereof — be they alternatives to classical logic, such as relevant logics, many-valued logics, intermediate logics, or extensions of classical logic, such as modal ($\square, \lozenge$) logics and temporal logics — do not fit well, from the proof-theoretic point of view, in Gentzen's original framework as characterized, in particular, by the adhesion to "standard" sequents $\Gamma \Rightarrow \Delta$, possibly with inessential variations (such as one-sided sequents) or restrictions (such as intuitionistic sequents) and to the coming in pairs (introduction in the antecedent and in the consequent) of logical rules.

In some cases, one may easily devise *ad hoc* Gentzen-style inference rules in order to capture this or that nonclassical logic, but the crucial problem is that these rules may not obey the subformula property or that the resulting calculus may not admit cut elimination.

Just to give an example, consider the minimal normal modal logic K (with $\square$ as primitive modal operator and $\lozenge\varphi$ defined as $\neg\square\neg\varphi$).[7] By adding to G3c$_\mathrm{p}$, the inference rule

$$\frac{\Gamma \Rightarrow \varphi}{\square\Gamma \Rightarrow \square\varphi, \Delta},$$

one obtains the sequent calculus GK, which turns out to be adequate for K. The above rule does not respect the pattern of Gentzen's logical rules (introduction in the antecedent or introduction in the consequent), yet GK is proof-theoretically well behaved, that is, it enjoys

---

[6]A sequent calculus for intuitionistic logic, named LJ, was introduced by Gentzen (1935) through a rather simple modification of LK, consisting of the restriction to sequents $\Gamma \Rightarrow \Delta$, where Δ contains at most one formula. LJ enjoys (syntactic) cut elimination, which entails the decidability of its propositional fragment LJ$_\mathrm{p}$. Sequent calculi for propositional intuitionistic logic that are more efficient than Gentzen's LJ$_\mathrm{p}$ from the point of view of proof-searching have been later devised and investigated.

[7]K is axiomatized by adding to an axiomatic calculus for classical propositional logic the axiom schema $\square(\varphi \to \psi) \to (\square\varphi \to \square\psi)$ and the necessitation rule $\dfrac{\varphi}{\square\varphi}$.

all the desirable properties: (syntactic) cut elimination and the sub-formula property. On the other hand, for many well-known extensions of K, among which is S5 (K + ($\Box\varphi \to \varphi$) + ($\Box\varphi \to \Box\Box\varphi$) + ($\varphi \to \Box\Diamond\varphi$)), there is no (and, in a precise sense, there cannot be any) adequate, more or less conventional sequent calculus.[8]

In his 2002 survey on the proof-theory of modal logics,[9] Wansing concluded that "no uniform way of presenting only the most important normal modal and temporal propositional logics as ordinary Gentzen calculi is known" and that, in any case, "the standard approach fails to be modular: in general it is not the case that a single axiom schema is captured by a single sequent rule (or a finite set of such rules)"— and we can say that this verdict also applies to other families of nonclassical logics.

Since the end of the 1980s, the proof-theoretical investigation of nonclassical logics has gradually led to the introduction of a variety of "nonconventional" Gentzen-type calculi or sequent-style calculi. The key point is the replacement of Gentzen's standard sequents with other kinds of related syntactic objects, which we may subsume under the category of "nonconventional sequents," such as the sequents adopted in "display" sequent calculi[10] (which feature, further to Gentzen's comma, a number of additional structural operators), the "hypersequents" (of the form $\Gamma_1 \Rightarrow \Delta_1 \mid \Gamma_2 \Rightarrow \Delta_2 \mid \cdots \mid \Gamma_n \Rightarrow \Delta_n$: finite lists, or sets, or multisets of conventional sequents),[11] the "nested" sequents[12] (see also the following section) and the "tree-hypersequents"[13] or the "labeled" sequents of Negri's labeled sequent calculi approach,[14] just to mention a few — see Wansing's 2002 chapter for a detailed survey.

---

[8] See Takano (2018).
[9] Wansing (2002).
[10] Belnap (1982).
[11] Avron (1991), Cf. also Avron (1996).
[12] Brünnler (2006). See also Brünnler and Strassburger (2009) and Goré *et al.* (2009).
[13] Poggiolesi (2011).
[14] Negri (2005).

This deviation from Gentzen's original format does not qualify a sequent-style calculus for a certain logic L as a "good" one. Typically, two major requirements have to be met, namely:

- the subformula property, and
- the cut-elimination theorem (better if is proved in a purely syntactical way),

which together pave the way to (more or less efficient) proof-searching algorithms. In certain cases, restricted exceptions to the subformula property or deviations from full cut elimination (the so-called "analytic cut elimination") may be tolerated.

In view of what we will see in the following section, we conclude with a concise presentation of the labeled sequent calculi approach developed by Negri, which has been applied successfully to modal and temporal logics and, in general, to (families of) logics which are semantically characterized by a Kripke-style, relational semantics.

In the nested sequents and the tree-hypersequents approaches, notions of the relational semantics underlying a given logic are implicitly internalized in the syntax (basically, the deep sequents and tree-hypersequents hide in a linear notation a tree-like structure whose nodes are standard sequents). The labeled sequent calculi approach is instead characterized by an explicit internalization of relational semantic notions within the syntax. Let us consider, by way of exemplification, the labeled, Gentzen-style calculus for the minimal normal modal logic K.

Besides the usual modal formulae (here, for convenience, a propositional language with $\bot, \wedge, \rightarrow, \Box, \Diamond$ as primitive logical operators is adopted), one needs variables $v_0, v_1, \ldots$ and a binary predicate letter R (the former, intuitively, ranging over the worlds or states of an unspecified Kripke model and the latter denoting its accessibility relation).

A relational atom has the form xRy, where x, y are variables. A labeled formula has the form $x : \varphi$, where x is a variable and φ is a formula (intuitive reading: φ is true at x). Finally, a labeled sequent has the standard form $\Gamma \Rightarrow \Delta$ but now Γ, Δ are finite multisets of labeled formulae and relational atoms.

The labeled calculus $G3_{lab}K$ is defined as follows.

Axioms (initial sequents):

$$x : p, \Gamma \Rightarrow \Delta, x : p, \qquad xRy, \Gamma \Rightarrow \Delta, xRy, \qquad x : \bot, \Gamma \Rightarrow \Delta;$$

Structural rule:

$$\frac{\Gamma \Rightarrow \Delta, x : \varphi \quad x : \varphi, \Pi \Rightarrow \Sigma}{\Gamma, \Pi \Rightarrow \Delta, \Sigma} \text{ cut;}$$

Propositional logical rules:

$$\frac{x : \varphi, x : \psi, \Gamma \Rightarrow \Delta}{x : \varphi \wedge \psi, \Gamma \Rightarrow \Delta} \wedge_l, \qquad \frac{\Gamma \Rightarrow \Delta, x : \varphi \quad \Gamma \Rightarrow \Delta, x : \psi}{\Gamma \Rightarrow \Delta, x : \varphi \wedge \psi} \wedge_r,$$

$$\frac{\Gamma \Rightarrow \Delta, x : \varphi \quad x : \psi, \Gamma \Rightarrow \Delta}{x : \varphi \rightarrow \psi, \Gamma \Rightarrow \Delta} \rightarrow_l, \qquad \frac{x : \varphi, \Gamma \Rightarrow \Delta, x : \psi}{\Gamma \Rightarrow \Delta, x : \varphi \rightarrow \psi} \rightarrow_r;$$

Modal logical rules:

$$\frac{y : \varphi, x : \Box\varphi, xRy, \Gamma \Rightarrow \Delta}{x : \Box\varphi, xRy, \Gamma \Rightarrow \Delta} \Box_l, \qquad \frac{xRy, \Gamma \Rightarrow \Delta, y : \varphi}{\Gamma \Rightarrow \Delta, x : \Box\varphi} \Box_r \text{ (!!}y\text{),}$$

$$\frac{xRy, y : \varphi, \Gamma \Rightarrow \Delta}{x : \Diamond\varphi, \Gamma \Rightarrow \Delta} \Diamond_l \text{ (!!}y\text{),} \qquad \frac{xRy, \Gamma \Rightarrow \Delta, x : \Diamond\varphi, y : \varphi}{xRy, \Gamma \Rightarrow \Delta, x : \Diamond\varphi} \Diamond_r,$$

with a proviso in the rules marked with (!!y): The variable y must not occur in the conclusion.

The axioms and the propositional logic rules need no special comments, considering that the semantics of the Boolean operators is local: They are just the corresponding axioms and rules of $G3c_p$, with the principal and secondary formula (or formulae) decorated with one and the same label.

The modal rules, which come in pairs and so fully respect Gentzen's pattern, are also natural since they can be easily justified by looking at the clauses:

$\mathcal{M}, x \vDash \Box\varphi$ if and only if $\mathcal{M}, y \vDash \varphi$ for all y such that xRy,
$\mathcal{M}, x \vDash \Diamond\varphi$ if and only if $\mathcal{M}, y \vDash \varphi$ for some y such that xRy,

defining the truth of a formula $\Box\varphi$ or $\Diamond\varphi$ at a world x in a model $\mathcal{M}$.[15]

$G3_{\text{lab}}K$ captures the modal system K (for all formulae φ: φ is a K-tautology iff the sequent $\Rightarrow x : \varphi$ is derivable in $G3_{\text{lab}}K$) and is indeed a good sequent-style calculus: It respects the (suitably reformulated) subformula principle and admits syntactical cut elimination. What is more important, $G3_{\text{lab}}K$ is the basic labeled modal calculus from which labeled sequent-style calculi for all the normal modal systems M, extending the minimal one K, can be modularly obtained, provided their corresponding frame condition in Kripke's semantics is expressible by universal formulae or geometric implications.[16] For such M's, the frame condition can be captured by suitable sequent-style so-called "mathematical rules" which, added to $G3_{\text{lab}}K$, give rise to a labeled sequent-style calculus $G3_{\text{lab}}M$ that is adequate for M and still enjoys the subformula property and cut elimination.

Just to give one example, the following mathematical rules

$$\frac{xRx, \Gamma \Rightarrow \Delta}{\Gamma \Rightarrow \Delta} \text{ Ref,} \qquad \frac{xRy, \Gamma \Rightarrow \Delta}{\Gamma \Rightarrow \Delta} \text{ Ser}(y!!),$$

$$\frac{xRy, yRx, \Gamma \Rightarrow \Delta}{xRy, \Gamma \Rightarrow \Delta} \text{ Sym,} \qquad \frac{xRz, xRy, yRz, \Gamma \Rightarrow \Delta}{xRy, yRz, \Gamma \Rightarrow \Delta} \text{ Trs}$$

---

[15]The soundness of the rules ($\Box_l$) and ($\Box_r$), and analogously that of ($\Diamond_l, \Diamond_r$), is proved as follows. Suppose the conclusion $x : \Box\varphi, xRy, \Gamma \Rightarrow \Delta$ of ($\Box_l$) is false in a Kripke model $\mathcal{M}$. Then, for the worlds x, y satisfying xRy, one has that $x \vDash \Box\varphi$ — so, by the semantics of $\Box$, $y \vDash \varphi$ — and also that each formula in Γ is true at some world (not necessarily the same), while each formula in Δ is false at some world (not necessarily the same). Hence, the premise $y : \varphi, x : \Box\varphi, xRy, \Gamma \Rightarrow \Delta$ of the rule is false in that Kripke model.

Next, suppose the conclusion $\Gamma \Rightarrow \Delta, x : \Box\varphi$ of ($\Box_r$) is false in a Kripke model $\mathcal{M}$. Then, each formula in Γ is true in some world (not necessarily the same), while each formula in Δ is false in some world (not necessarily the same), and $\Box\varphi$ is false in the world x. This implies that there must be a world accessible from x, in which φ is false. We are not entitled to make additional assumptions on this world, so we pick one such y not occurring in the conclusion. Hence, the premise $xRy, \Gamma \Rightarrow \Delta, y : \varphi$ of the rule is false in that Kripke model.

[16]A universal formula is a closed formula of the form $\forall x_1 \ldots \forall x_n A$, where A is quantifier-free. A geometric implication is a formula of the form $\forall x_1 \ldots \forall x_n (A \rightarrow B)$, where A and B are $\rightarrow$- and $\forall$-free.

correspond, respectively, to the frame conditions of reflexivity, seriality, symmetry, and transitivity. Thus, for example, the labeled calculus $G3_{lab}S5$ defined as $G3_{lab}K+\{(Ref), (Sym), (Trs)\}$ provides a good sequent-style calculus for the modal system S5.

3.1.3.   Sequent-style calculi for TL

To the best of our knowledge, the first systematic attempt to devise a sequent calculus for Prior's basic system TL of temporal logic is due to Nishimura in 1980.[17]

The calculus GK_t he introduces works with standard, set-based, sequents, thus the only structural rules are Weakening and the cut rule. The logical rules for the Boolean operators (primitive connectives: negation and conditional) are those of LK_p. The logical rules for the primitive temporal operators H and G (P and F are taken as defined) are two right-introduction rules, not accompanied by corresponding left-introduction rules[18]:

$$\frac{\Gamma \Rightarrow H\Delta, \varphi}{G\Gamma \Rightarrow \Delta, G\varphi} \ G_r, \qquad \frac{\Gamma \Rightarrow G\Delta, \varphi}{H\Gamma \Rightarrow \Delta, H\varphi} \ H_r.$$

One can easily see that these rules are sound with respect to the Kripke semantics for TL, i.e., they preserve truth in an arbitrary model. Suppose indeed that the conclusion of $[G_r]$ is false in a TL-model $\mathcal{M}$. Then, there is an instant i in $\mathcal{M}$ at which all formulae in $G\Gamma$ are true and all formulae in $\Delta \cup \{G\varphi\}$ are false. By $i \nvDash G\varphi$, it follows that $j \nvDash \varphi$ for some $j > i$. But then also, all formulae in $H\Delta$ are false at j, whereas all formulae in Γ are true at j. Hence the premise of $[G_r]$ being false in $\mathcal{M}$. The soundness of $[H_r]$ is verified analogously.

Thus, GK_t is sound with respect to TL-validity. Nishimura proves that it is also complete and so equivalent to the axiomatic characterization of TL.

GK_t is not, however, a good sequent calculus for TL. First, both the rules $[G_r]$ and $[H_r]$ patently violate the subformula principle. Next, the cut rule cannot be eliminated. For example, it is not difficult to realize that the characteristic TL-tautology $\varphi \rightarrow G\neg H\neg\varphi$

---

[17]Nishimura (1980).
[18]$G\Gamma$ is an abbreviation for $\{G\varphi \mid \varphi \in \Gamma\}$ and, analogously, $H\Gamma$.

(i.e., $\varphi \to \mathrm{GP}\varphi$) cannot be derived in $\mathrm{GK_t}$ without the application of the cut rule.

By suitably modifying the rules $[\mathrm{G_r}]$ and $[\mathrm{H_r}]$, Nishimura also provides a Gentzen-style formulation $\mathrm{GK4_t}$ for TL + Transitivity that shares, however, all the negative aspects of $\mathrm{GK_t}$.

To conclude, it is perhaps interesting to recall that Nishimura also introduced in the same paper a second sequent calculus $\mathrm{GHK_t}$ provably sound and complete with respect to TL-validity. It is based on nonconventional sequents (a kind of forerunners of hypersequents) of the form

$$\Gamma_1; \Gamma_2; \Gamma_3 \Rightarrow \Delta_1; \Delta_2; \Delta_3$$

(the Γ's and Δ's are finite sets of formulae; the ";" is a new structural operator), intuitively corresponding to the standard sequent

$$\mathrm{H}\Gamma_1, \Gamma_2, \mathrm{G}\Gamma_3 \Rightarrow \mathrm{H}\Delta_1, \Delta_2, G\Delta_3$$

and so having the intended meaning of the temporal formula

$$\bigwedge \mathrm{H}\Gamma_1 \wedge \bigwedge \Gamma_2 \wedge \bigwedge \mathrm{G}\Gamma_3 \to \bigvee \mathrm{H}\Delta_1 \vee \bigvee \Delta_2 \vee \bigvee G\Delta_3.$$

Unlike $\mathrm{GK_t}$, all the rules of $\mathrm{GHK_t}$ now respect the subformula principle. But unfortunately, the cut rule cannot be dispensed with, as in $\mathrm{GK_t}$.

The sequent-style calculi for TL and some of its extensions (TL plus one of Transitivity, Reflexivity, Connectedness, or combinations thereof) introduced by Kashima in 1994[19] are instead examples (the first ones, perhaps) of good sequent-style calculi for temporal logic:

- they enjoy cut elimination (although the given proof of the result is only semantic and not syntactic), and
- all the other inference rules are analytic, i.e., they fully respect the subformula principle.

The price to pay is to abandon Gentzen's standard sequents and to work instead with a peculiar elaboration of the Tait-style approach (see the previous section).

---

[19] Kashima (1994).

First of all, the propositional TL language is recasted in a convenient form featuring, besides the literals and the Boolean operators $\wedge, \vee$, all the four Prior's temporal operators H, G, P, F, and the definition of $\neg\varphi$ is extended by the natural clauses: $\neg H\varphi := P\neg\varphi, \neg G\varphi := F\neg\varphi, \neg P\varphi := H\neg\varphi, \neg F\varphi := G\neg\varphi$.

Next, nonstandard sequents are adopted, namely one-sided, nested sequents. More precisely, the class $\mathcal{S}$ of Kashima sequents (henceforth, sequents), denoted by $S, T, \ldots$, is defined inductively as the smallest class such that:

— $\varphi \in \mathcal{S}$, for any formula φ;

— if $S \in \mathcal{S}$, then $[S]^P, [S]^f \in \mathcal{S}$;

— if $\{S_i\}_{1\leq i\leq n} \subseteq \mathcal{S}$ $(n \geq 0)$, then $S_1, S_2, \ldots, S_n \in \mathcal{S}$
  (if $n = 0$, this is the empty sequent).

The meaning of a sequent S is given by the temporal formula S^* inductively associated to it as follows:

— if $S \equiv \varphi$, $S^* := \varphi$;

— $([S]^P)^* := HS^*$ and $([S]^f)^* := GS^*$;

— $(S_1, \ldots, S_n)^* := S_1^* \vee \ldots \vee S_n^*$ ($\bot$ if $n = 0$).

Kashima's basic sequent system SK_t, corresponding to TL, is now obtained by adding to the Tait-style sequent calculus for LK_p previously described the following

Structural rules:

$$\frac{\Gamma, [\Delta]^f}{[\Gamma]^P, \Delta} \text{ turn}_1, \qquad \frac{\Gamma, [\Delta]^P}{[\Gamma]^f, \Delta} \text{ turn}_2$$

and

Temporal inference rules:

$$\frac{\Gamma, [\varphi]^P}{\Gamma, H\varphi} H, \qquad \frac{\Gamma, [\varphi]^f}{\Gamma, G\varphi} G, \qquad \frac{\Gamma, [\Delta, \varphi]^P}{\Gamma, [\Delta]^P, P\varphi} P, \qquad \frac{\Gamma, [\Delta, \varphi]^f}{\Gamma, [\Delta]^f, F\varphi} F.$$

To exemplify the use of Kashima's temporal rules, here is a SK_t-derivation of the formula (one element sequent) $\neg p \vee HFp$, alias is the TL-tautology $p \to \mathrm{HF}p$:

$$\cfrac{\cfrac{\cfrac{\cfrac{\cfrac{\neg p, p}{[\,\neg p, p\,]^{\mathrm{f}}}\ \mathrm{turn}_2}{\mathrm{F}p, [\,\neg p\,]^{\mathrm{f}}}\ \mathrm{F}}{\neg p, [\,\mathrm{F}p\,]^{\mathrm{P}}}\ \mathrm{turn}_1.}{\neg p, \mathrm{HF}p}\ \mathrm{H}}{\neg p \vee \mathrm{HF}p}\ \vee$$

Truth preservation of the temporal rules in an arbitrary TL model can indeed be easily verified, and the soundness of Kashima's calculus SK_t follows. Completeness, together with (semantic) cut elimination, is proved by Kashima through a suitable adaptation of the known canonical-tree construction technique. Given an input sequent S, the associated search tree is proved to yield either a cut-free derivation of S in SK_t or a TL countermodel for S*.

We conclude this section with a presentation of the labeled sequent-style calculus $\mathrm{G3_{lab}K_t}$ for the basic temporal system TL and of its modular extensions corresponding to extensions of TL derived from the addition of certain conditions on the temporal precedence relation.[20]

The labeled-sequents framework for plain modal logics has already been already described, and its adaptation to basic temporal logics is quite natural.

The propositional language features as primitives the propositional constants $\bot$ (absurd), $\wedge$ (conjunction), $\to$ (conditional), and Prior's temporal operators $\mathrm{H, G, P, F}$; $\neg\phi$ is defined as $\varphi \to \bot$.

The definitions of relational atoms, labeled formulae, and labeled sequents remain unchanged, except that for obvious reasons, the relational symbol "R" is replaced by "$<$."

$\mathrm{G3_{lab}K_t}$ is now defined as follows:

Axioms (initial sequents):

$$x : p, \Gamma \Rightarrow \Delta, x : p, \qquad x < y, \Gamma \Rightarrow \Delta, x < y, \qquad x : \bot, \Gamma \Rightarrow \Delta.$$

Structural rule (cut) and propositional logic rules: as in $\mathrm{G3_{lab}K}$.

---

[20]Boretti (2008).

Temporal logic rules:

$$\frac{y : \varphi, x : H\varphi, y < x, \Gamma \Rightarrow \Delta}{x : H\varphi, y < x, \Gamma \Rightarrow \Delta} \; H_1, \qquad \frac{y < x, \Gamma \Rightarrow \Delta, y : \varphi}{\Gamma \Rightarrow \Delta, x : H\varphi} \; H_r \, (!!y),$$

$$\frac{y : \varphi, x : G\varphi, x < y, \Gamma \Rightarrow \Delta}{x : G\varphi, x < y, \Gamma \Rightarrow \Delta} \; G_1, \qquad \frac{x < y, \Gamma \Rightarrow \Delta, y : \varphi}{\Gamma \Rightarrow \Delta, x : G\varphi} \; G_r \, (!!y),$$

$$\frac{y < x, y : \varphi, \Gamma \Rightarrow \Delta}{x : P\varphi, \Gamma \Rightarrow \Delta} \; P_1 \, (!!y), \qquad \frac{y < x, \Gamma \Rightarrow \Delta, x : P\varphi, y : \varphi}{y < x, \Gamma \Rightarrow \Delta, x : P\varphi} \; P_r,$$

$$\frac{x < y, y : \varphi, \Gamma \Rightarrow \Delta}{x : F\varphi, \Gamma \Rightarrow \Delta} \; F_1 \, (!!), \qquad \frac{x < y, \Gamma \Rightarrow \Delta, x : F\varphi, y : \varphi}{x < y, \Gamma \Rightarrow \Delta, x : F\varphi} \; F_r.$$

Just to give an example of how the calculus works, here is a derivation in $G3_{lab}K_t$ of the labeled sequent $x : \varphi \to HF\varphi$ (corresponding to the TL tautology $\varphi \to HF\varphi$):

$$\frac{\dfrac{y < x, x : \varphi \Rightarrow y : F\varphi, x : \varphi}{\dfrac{y < x, x : \varphi \Rightarrow y : F\varphi}{\dfrac{x : \varphi \Rightarrow x : HF\varphi}{\Rightarrow x : \varphi \to HF\varphi} \to_r.}{} H_r} F_r$$

One can indeed prove that the calculus $G3_{lab}K_t$ is indeed, like $G3_{lab}K$, a good sequent-style calculus for TL.

Now, $G3_{lab}K_t$ can be extended modularly and preserving all the "good" properties by adding mathematical rules corresponding to possible conditions on the temporal precedence relation $<$ that are expressible by universal formulae or geometric implications.

Just to give an idea of this method, consider, for instance (further to the examples given for a generic R in the previous section), the following four conditions on $<$ (the first two are expressed by universal formulae, while the third and the fourth by geometric implications):

- Left linearity: $\forall xyz(y < x \wedge z < x \to y < z \vee y = z \vee z < y)$;
- Right linearity: $\forall xyz(x < y \wedge x < z \to y < z \vee y = z \vee z < y)$;
- Density: $\forall xy(x < y \to \exists z(x < z \wedge z < y))$;
- Right directedness: $\forall xyz(x < y \wedge x < z \to \exists w(y < w \wedge z < w))$.

Note that, further to the relational symbol $<$, also the relation $=$ of equality occurs in three of the above four conditions. Accordingly, one has to first generalize the notion of relational atom by including also relational atoms of the form $x = y$. At the same time, suitable mathematical rules for equality have to be added to the calculus $G3_{lab}K_t$ (in the following, At denotes a relational atom):

$$\frac{x = x, \Gamma \Rightarrow \Delta}{\Gamma \Rightarrow \Delta} \ \text{EqRef},$$

$$\frac{y : p, x = y, x : p, \Gamma \Rightarrow \Delta}{x = y, x : p, \Gamma \Rightarrow \Delta} \ \text{EqSub}_1, \qquad \frac{\text{At}[x/y], x = y, \text{At}, \Gamma \Rightarrow \Delta}{x = y, \text{At}, \Gamma \Rightarrow \Delta} \ \text{EqSub}_2.$$

Finally, one adds to the calculus one (or a combination) of the mathematical rules corresponding to the four properties above:

$$\frac{y < z, y < x, z < x, \Gamma \Rightarrow \Delta \quad y = z, y < x, z < x, \Gamma \Rightarrow \Delta \quad z < y, y < x, z < x, \Gamma \Rightarrow \Delta}{y < x, z < x, \Gamma \Rightarrow \Delta} \ \text{L-Lin},$$

$$\frac{y < z, x < y, x < z, \Gamma \Rightarrow \Delta \quad y = z, x < y, x < z, \Gamma \Rightarrow \Delta \quad z < y, x < y, x < z, \Gamma \Rightarrow \Delta}{x < y, x < z, \Gamma \Rightarrow \Delta} \ \text{R-Lin},$$

$$\frac{x < z, z < y, x < y, \Gamma \Rightarrow \Delta}{x < y, \Gamma \Rightarrow \Delta} \ \text{Dens (!!}z), \qquad \frac{y < u, z < u, x < y, x < z, \Gamma \Rightarrow \Delta}{x < y, x < z, \Gamma \Rightarrow \Delta} \ \text{R-Dir (!!}u).$$

All the temporal sequent-style calculi obtained in this way admit syntactic cut elimination.

3.1.4. *Sequent-style calculi for* **LTL** *and* **CTL**: *A challenge*

One of the distinctive traits of LTL (and of course also of CTL and CTL*) with respect to TL, among many others, concerns already Prior's basic temporal operators G and F, which in these systems acquire a precise infinitary meaning (in other words, they provide the greatest, resp. the least, fixpoint of suitable operators), i.e., a meaning which is intuitively expressed by the infinitary conjunction, resp. disjunction:

$$(\text{G}\varphi): \ [\varphi \wedge] \text{X}\varphi \wedge \text{X}^2\varphi \wedge \cdots \wedge \text{X}^n\varphi \wedge \cdots,$$

$$(\text{P}\varphi): \ [\varphi \vee] \text{X}\varphi \vee \text{X}^2\varphi \vee \cdots \vee \text{X}^n\varphi \vee \cdots,$$

where X is the *next*-operator and $X^n \varphi$ $(n \geq 0)$ is an abbreviation for $\overbrace{X \cdots X}^{n \text{ times}} \varphi$.[21]

This peculiarity has naturally led the initial search for sequent-style calculi for these logics in the direction either of sequent-style infinitary calculi (i.e., calculi featuring some rule with a countably infinite number of premises) or of finitary calculi that, even though admitting cut elimination, do not obey the subformula principle because they contain some rules (the so-called invariant rules, which exploit the greatest fixed point nature of G and U) that violate the subformula property.

Let us illustrate each of these two different approaches with a representative example. Preliminarily, note that due to the absence in LTL of temporal operators looking to the past, there is no special need to use nonstandard sequents (see the problem previously discussed for TL).

The propositional fragment LT_ω of the infinitary sequent calculus for the quantified version of LTL without the Until operator, introduced by Kaway in 1987,[22] is defined as follows.

The sequents are (as in Gentzen's original calculus LK_p) of the form $\Gamma \Rightarrow \Delta$, where Γ, Δ are finite, possibly empty lists of formulae of the propositional language, which contains the primitive connectives $\neg, \wedge$ and the temporal operators X, G, F.

The initial sequents and the structural rules are exactly those of Gentzen's LK_p. The remaining rules are as follows, for each $k, i \geq 0$:

---

[21] In LTL and its extensions, one can take G as a primitive operator together with the Until operator U, or define it by means of U: $G\varphi := \neg(\top U \neg \varphi)$. Most authors adopt for G and U a "wide" semantic evaluation clause, according to which, for example, $G\varphi$ is true at an instant $n \in \omega$ iff φ is true at all instants $m \geq n$ (and dually for $F\varphi$), while others keep instead for G the evaluation clause of TL that excludes the present instant. This is the reason why the first conjunct (disjunct) φ in the above infinitary formulae appears in square brackets. The choice between the wide or the strict (present instant excluded) truth condition is in any case a matter of preference, that is, it has no influence at all on the problem we are discussing.

[22] Kawai (1987).

Logical rules:

$$\frac{\Gamma \Rightarrow \Delta, \mathrm{X}^k \varphi}{\mathrm{X}^k \neg \varphi, \Gamma \Rightarrow \Delta} \neg_1^k, \qquad \frac{\mathrm{X}^k \varphi, \Gamma \Rightarrow \Delta}{\Gamma \Rightarrow \Delta, \mathrm{X}^k \neg \varphi} \neg_r^k,$$

$$\frac{\mathrm{X}^k \varphi_i, \Gamma \Rightarrow \Delta}{\mathrm{X}^k (\varphi_1 \wedge \varphi_2), \Gamma \Rightarrow \Delta} \wedge_1^k (i=1,2), \qquad \frac{\Gamma \Rightarrow \Delta, \mathrm{X}^k \varphi \quad \Gamma \Rightarrow \Delta, \mathrm{X}^k \psi}{\Gamma \Rightarrow \Delta, \mathrm{X}^k (\varphi \wedge \psi)} \wedge_r^k.$$

Rules for G and F:

$$\frac{\mathrm{X}^{k+i} \varphi, \Gamma \Rightarrow \Delta}{\mathrm{X}^k \mathrm{G} \varphi, \Gamma \Rightarrow \Delta} \mathrm{G}_1^{k,i}, \qquad \frac{\{\cdots \Gamma \Rightarrow \Delta, \mathrm{X}^{k+j} \varphi \cdots\}_{j \in \omega}}{\Gamma \Rightarrow \Delta, \mathrm{X}^k \mathrm{G} \varphi} \mathrm{G}_r^k,$$

$$\frac{\{\cdots \mathrm{X}^{k+j} \varphi, \Gamma \Rightarrow \Delta \cdots\}_{j \in \omega}}{\mathrm{X}^k \mathrm{F} \varphi, \Gamma \Rightarrow \Delta} \mathrm{F}_1^k, \qquad \frac{\Gamma \Rightarrow \Delta, \mathrm{X}^{k+i} \varphi}{\Gamma \Rightarrow \Delta, \mathrm{X}^k \mathrm{F} \varphi} \mathrm{F}_r^{k,i}.$$

Given the intended semantics, the inference rules are quite natural, and their soundness can be easily verified.[23]

Note that the inference rules satisfy a weak version of the subformula property (one has to relax the notion of subformula by counting, for example, $\mathrm{X}^k \varphi$ among the subformulae of $\mathrm{X}^k \neg \varphi$, etc.) and, most importantly, that the rules (G_r^k) and (F_1^k) have an infinite number of premises.

Kaway's LT_ω provably captures the U-free fragment of linear-time temporal logic LTL and admits cut elimination. Other infinitary sequent-style systems for LTL or systems in the neighborhood have been devised, for example, by Baratella and Masini[24] and, in the framework of labeled sequent calculi, by Masini and Negri.[25]

---

[23]For instance, suppose the conclusion of (G_r^k) is false in a LTL-model $\mathcal{M}$. Then, all the Γ's (Δ's) are true (false) at the initial position 0, and also $\mathrm{X}^k \mathrm{G} \varphi$ is false at 0. Thus, $\mathrm{G} \varphi$ must be false at k; therefore, φ must be false at $k+j$ for some $j \geq 0$. Therefore, $\mathrm{X}^{k+j} \varphi$ is false at the initial position 0, so at least one of the infinitely many premises of the rule is false in $\mathcal{M}$.

[24]Baratella and Masini (2004). Cf. also Kamide (2006)

[25]Negri and Boretti (2010). In this paper the authors introduce and investigate an interesting infinitary labeled sequent-style calculus for Prior's "system 7.2" (Prior, 1967, p. 178), which is characterized by the presence of both future and past operators: X, Y (yesterday), G, and H. The intended underlying time flow has the order-type $\omega^* + \omega$ of the integers $\langle Z, < \rangle$.

The sequent calculus G_{LTL} for full LTL introduced by Paech[26] is instead finitary and admits cut elimination, but it contains an invariant rule.

G_{LTL} features standard, set-based sequents $\Gamma \Rightarrow \Delta$ over a propositional language containing $\neg, \vee, X$, and W as primitive Boolean, resp. temporal, operators. Here, W is the so-called "weak until" (or "unless") operator, whose semantics is that of the (strong) until operator U but without the stop condition (so $\varphi W \psi$ is equivalent to $G\varphi \vee \varphi U \psi$).

The initial sequents and the structural rules (Weakening and cut only since the antecedent and the consequent of a sequent are sets of formulae) of G_{LTL} are those of Gentzen's LK_p, as well as the introduction rules for $\neg$ and $\vee$ (except that $(\vee_r)$ is taken in the additive formulation). Finally, the introduction rules for X and W are as follows, where $X\Gamma := \{X\varphi \mid \varphi \in \Gamma\}$ and $^\vee\Delta := \bigvee\{\varphi \mid \varphi \in \Delta\}$:

$$\frac{\Gamma \Rightarrow \Delta}{X\Gamma \Rightarrow X^\vee\Delta}\ X, \qquad \frac{X\neg\varphi, \Gamma \Rightarrow \Delta}{\neg X\varphi, \Gamma \Rightarrow \Delta}\ \neg X_l, \qquad \frac{\Gamma \Rightarrow \Delta, \neg X\varphi}{X\Gamma \Rightarrow \Delta, X\neg\varphi}\ X\neg_r,$$

$$\frac{X(\psi \vee (\varphi \wedge \varphi W\psi)), \Gamma \Rightarrow \Delta}{\varphi W\psi, \Gamma \Rightarrow \Delta}\ W_l, \qquad \frac{\Gamma \Rightarrow \Delta, X(\psi \vee (\varphi \wedge \varphi W\psi))}{\Gamma \Rightarrow \Delta, \varphi W\psi}\ W_r$$

$$\frac{\Gamma \Rightarrow \Delta, \chi \qquad \chi \Rightarrow X(\psi \vee (\varphi \wedge \chi))}{\Gamma \Rightarrow \Delta, \varphi W\psi}\ W_{Ind}, \qquad \text{where } \chi \text{ is a } W\text{-}formula.$$

The three rules for X are quite natural and are easily seen to be sound. The two rules (W_l) and (W_r) correspond to the LTL's fixpoint equivalence

$$\varphi W\psi \leftrightarrow \psi \vee (\varphi \wedge X(\varphi W\psi))$$

holding for the operator W, while the induction rule (W_{Ind}) is necessary to capture the greatest fixpoint operator character of W. This rule, however, is an example of an invariant rule which, due to the presence of the W-formula χ in the premise, clearly violates the subformula property — in other words, a cut restricted to formulae of the form $\chi_1 W\chi_2$ is involved in this rule. From the point of view of

---

[26]Paech (1988).

proof-searching, the presence of this rule sensibly reduces the advantages given by the cut-elimination theorem for G_{LTL}.

The challenge of devising finitary, cut-free, and invariant rules-free sequent-style calculi for LTL and its extensions CTL and CTL* is still in the agenda of proof-theoretical investigations.[27] As far as we know, up to now, there are no significative results for the strongest system CTL* in the literature. Some promising results have instead been obtained for the two weaker systems, in particular those contained in two (independent) papers by Brünnler and Lange,[28] and by Gaintzarain *et al.*[29]

The sequent-style calculi for both LTL and CTL introduced in the former paper employ a particularly complex formalism based on annotating fixpoint formulae with a "history" (a finite set of finite sets of formulae) and cannot be easily illustrated here.

We thus conclude with the presentation of the finitary, cut-free, and invariant rules-free sequent-style calculus FC for linear-time temporal logic LTL introduced in the second paper.

The primitive logical symbols of the propositional language of FC are the Boolean connectives $\neg$ and $\vee$ together with the propositional constant $\bot$ and the temporal operators X and U. The format for the sequents is set-based and single-consequent, that is, a sequent has the form

$$\Gamma \Rightarrow \varphi,$$

where Γ is a finite, possibly empty, set of formulae and φ is a formula.

The sequent calculus FC is construed as follows:

Initial sequents: $\Gamma, \varphi \Rightarrow \varphi$.

Structural rules:

$$\frac{\Gamma \Rightarrow \varphi}{\Gamma, \Delta \Rightarrow \varphi} \text{ W},
\qquad
\frac{\Gamma, \neg\varphi \Rightarrow \bot}{\Gamma \Rightarrow \varphi} \text{ dn},
\qquad
\frac{\Gamma \Rightarrow X\bot}{\Gamma \Rightarrow \psi} \text{ X}\bot.$$

---

[27] An analogous challenge concerns other modal logics which contain fixpoint operators, e.g., the common knowledge logic, the dynamic modal logic, and, more generally, the μ-calculus.

[28] Brünnler and Lange (2008).

[29] Gaintzarain *et al.* (2007).

Introduction rules for $\neg$ and $\vee$:

$$\frac{\Gamma \Rightarrow \varphi}{\Gamma, \neg\varphi \Rightarrow \psi} \, \neg_l, \qquad\qquad \frac{\Gamma, \varphi \Rightarrow \bot}{\Gamma \Rightarrow \neg\varphi} \, \neg_r,$$

$$\frac{\varphi, \Gamma \Rightarrow \chi \quad \psi, \Gamma \Rightarrow \chi}{\varphi \vee \psi, \Gamma \Rightarrow \chi} \, \vee_l, \qquad\qquad \frac{\Gamma \Rightarrow \varphi_i}{\Gamma \Rightarrow \varphi_1 \vee \varphi_2} \, \vee_r \, (i=1,2).$$

Introduction rules for X:

$$\frac{X\Gamma \Rightarrow \varphi}{\Gamma \Rightarrow X\varphi} \, X, \qquad \frac{\Gamma, X\neg\varphi \Rightarrow \psi}{\neg X\varphi, \Gamma \Rightarrow \psi} \, \neg X_l, \qquad \frac{\Gamma \Rightarrow \neg X\varphi}{\Gamma \Rightarrow X\neg\varphi} \, X\neg_r.$$

Introduction rules for U:

$$\frac{\Gamma, \psi \Rightarrow \chi \quad \Gamma, \varphi, \neg\psi, X(\vartheta_i U\psi) \Rightarrow \chi}{\Gamma, \varphi U\psi \Rightarrow \chi} \, U_l \, (i=1,2)$$

(where $\vartheta_1 := \varphi$ and $\vartheta_2 := \varphi \wedge (\Gamma^\neg \vee \chi)$, with $\Gamma^\neg := \bigvee\{\neg\xi_i \mid \xi_i \in \Gamma\}$),

$$\frac{\Gamma, \neg\varphi \Rightarrow \psi \quad \Gamma, \varphi, \neg X(\varphi U\psi) \Rightarrow \psi}{\Gamma \Rightarrow \varphi U\psi} \, U_r.$$

The cut rule is not included in the calculus: The authors prove indeed the completeness of FC with respect to LTL's semantics and so, indirectly, the admissibility of the cut rule. Since all the rules are finitary, and there are no invariant rules with hidden cuts, one may say that FC meets the requirements of a well-behaved sequent-style calculus for LTL.

The calculus is not, however, optimal, nor particularly elegant. Note, for example, that the choice of single-consequent sequents (which, as we said, are tailored for intuitionistic logic) forces the inclusion of the rule (dn) among the structural rules. And note also that a number of rules satisfy the subformula property only with respect to a very loose notion of subformula, a fact which makes the root-first, bottom-up proof-searching strategy quite inefficient.

3.2. Tableaux-Based Calculus and Temporal Logic

The main decision problems for temporal logic are satisfiability, validity, and model-checking. In order to check a formula φ for validity, one can check $\neg\varphi$ for satisfiability and invert the result because φ is valid iff $\neg\varphi$ is unsatisfiable. Satisfiability can be reduced to validity.

Satisfiability-checking procedures can be realized by interpreted transition systems (ITSs).

3.2.1.   *What are tableaux?*

In classical logic, a systematic search for a model of a formula can be realized by a tree-like graph, which is called tableau. A tableau search builds a graph step by step, following the syntactic structure of the formula and unfolding it into simpler formulae, until all requirements corresponding to the truth of the formulae in the labels are fulfilled. This method is similar to the proof search procedures in Gentzen-style deductive systems because they also build proof trees. Historically, the tableau procedure dates back to the Dutch logician E. W. Beth (1908–1964), who noted the link between Gentzen's work and his tableaux.[30]

Tableaux of temporal formulae may arrive at nodes which carry the same list of formulae in the label as nodes encountered before. For decision problems, such as satisfiability, the search procedure of a model must be terminated. Therefore, the notions of closed and open tableaux must be defined.

In linear temporal logic (LTL), the tableau graph of formula $GFp \land GF\neg p$ can be developed step by step.[31] The tableau grows as a directed graph, following the rules for generating new nodes from existing ones. The nodes are sets of formulae. The generating rules follow the formula decomposition rules for the Boolean connectives, temporal operators, and a rule for creating a successor of the current state with a label collecting all X-prefixed subformulae in the label of that state.

When a state with an already existing label is generated, the graph loops back to the already existing state with that label. When a label with a contradictory pair of formulae is generated, the branch terminates with a label $\bot$. The tableau in Fig. 3.1 demonstrates that the input formula is satisfiable and delivers a satisfying linear model.

If the search for a satisfying model terminates without success, the tableau is pronounced "closed," and the input set of formulae is declared unsatisfiable. If the tableau method is sound, closure of the

---

[30]Beth (1955), De Jongh (2008).
[31]Wolper (1985).

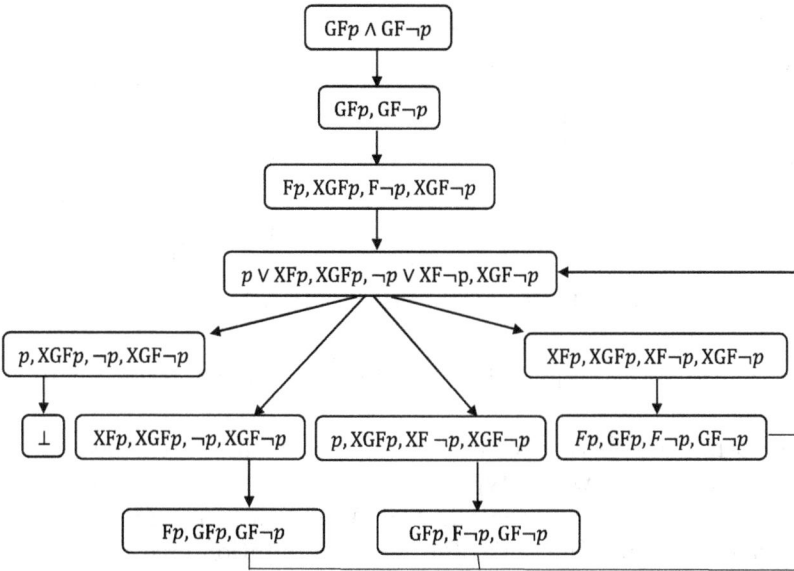

Fig. 3.1.   Tableau of a satisfiable LTL formula.[32]

tableau search must imply that the input set of formulae is unsatisfiable. If the search succeeds or never terminates, the tableau is called "open." If the input is not satisfiable, a complete tableau procedure is supposed to terminate.

3.2.2.   Basics of tableau construction in basic modal logic[33]

The set of all subformulae of a formula η is denoted by $sub(\eta)$, resp. the set $sub(\Gamma)$, of all subformulae of a set Γ of formulae. Basic logical connectives in basic modal logic (BML) are $\top, \bot, \neg, \wedge, \vee, \to$, EX, and AX.

There are four types of formulae (Fig. 3.2):

(1)   A conjunctive formula is associated with the set of its conjunctive components. For example, the conjunctive components of $\varphi \wedge \psi$ are φ and ψ. The only conjunctive component of $\neg\varphi$ is φ.

---

[32]Demri *et al.* (2016, p. 469).
[33]Tableaux were adapted for modal logic by, for example, Kripke (1963).

Conjunctive		Disjunctive		Successor	
Formula	Components	Formula	Components	Formula	Components
$\neg\neg\varphi$	φ, φ	$\varphi \vee \psi$	φ, ψ	$EX\varphi$	φ
$\varphi \wedge \psi$	φ, ψ	$\varphi \rightarrow \psi$	$\neg\varphi, \psi$	$AX\varphi$	φ
$\neg(\varphi \vee \psi)$	$\neg\varphi, \neg\psi$	$\neg(\varphi \wedge \psi)$	$\neg\varphi, \neg\psi$	$\neg EX\varphi$	$\neg\varphi$
$\neg(\varphi \rightarrow \psi)$	$\varphi, \neg\psi$			$\neg AX\varphi$	$\neg\varphi$

Fig. 3.2.   Types and components of formulae in BML.[34]

(2) A disjunctive formula is associated with the set of its disjunctive components.

(3) Formulae referring to truth in successor states are called successor formulae, which typically have the form $O\varphi$ or $\neg O\varphi$, with modal operator O for EX and AX in case of BML, temporal logic of reachability (TLR), and computation tree logic (CTL) and X in LTL. The only successor component of $O\varphi$ is φ, resp. $\neg\varphi$, for $\neg O\varphi$. In general, the successor component of a formula η is denoted by $scomp(\eta)$.

(4) Literals are $\top$, $\bot$, atomic propositions, and negations of these.

The (extended) closure $ecl(\varphi)$ of the formula φ is the least set of formulae such that

(1) $\varphi \in ecl(\varphi)$,

(2) $ecl(\varphi)$ is closed under all conjunctive, disjunctive, and successor components of the respective formulae in $ecl(\varphi)$.[35]

Example: $ecl(\varphi) = \{\varphi, AXp, p, \neg AXAXp, \neg AXp, \neg p\}$.

In classical propositional logic (PL), the formula decomposition rules of tableau construction can be extracted from the truth tables of the propositional connectives (Fig. 3.3).

The tableau construction in PL is a decomposition of an input formula applying these rules. Step by step, a tree of nodes labeled by sets of formulae is generated until a stage of saturation is reached. In this case, no new formula can be produced on any branch of the tree

---

[34]Demri *et al.* (2016, p. 480).
[35]Fischer and Ladner (1979).

Non-branching rules   Branching rules

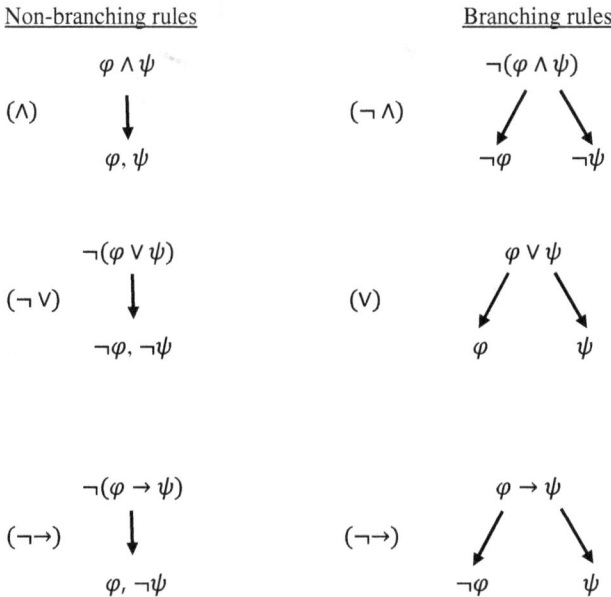

Fig. 3.3.   Branching and non-branching rules of tableaux in propositional logic.[36]

by applying a rule to any of the formulae in the labels of the nodes on that branch.

A branch is closed if it ends with either of $\top$, $\neg\bot$, or a pair of φ and $\neg\varphi$. Otherwise, the branch is open. A saturated tableau is called open if it has at least one open branch, otherwise it is called closed. The tableau system for PL is closed iff the input formula is not satisfiable. For temporal formulae, the tableau of PL must be extended with rules for temporal conjunctive and disjunctive formulae or their negations.

A set of formulae is called patently inconsistent if it contains $\top$, or $\neg\bot$, or a pair of φ and $\neg\varphi$. A set Γ of formulae is fully expanded iff (1) it is not patently inconsistent, (2) for every conjunctive formula in Γ, all of its conjunctive components are in Γ, and (3) for every disjunctive formula in Γ, at least one of its disjunctive components is in Γ.

---

[36]Demri *et al.* (2016, p. 482).

In classical PL, every open branch in a saturated tableau delivers at least one satisfying truth assignment. It follows that every fully expanded set of propositional formulae is satisfiable. In temporal logic, not every fully expanded set is satisfiable because a contradiction may occur not only at the current state ("locally") but elsewhere in the model. Therefore, in temporal logic, it must be checked whether at least one full expansion of the input formula set is satisfiable. An example is the formula $\eta = \neg(\mathrm{AX}p \to \mathrm{AXAX}p)$, which only has one full expansion $\{\eta, \mathrm{AX}p, \neg\mathrm{AXAX}p\}$.

An ITS determines the truth value of every formula at every state. We are interested in an information system that only contains just enough information to determine the truth values of those formulae that are directly concerned by the evaluation of the input formula at the root of a tableau. A so-called Hintikka structure is a partly defined ITS satisfying the input formula. It is represented by a graph of nodes labeled by a set of formulae. These labels are fully expanded subsets of the extended closure of a designated input formula. The labeling of states must ensure that the Hintikka structure can generate a model of the input formula.

It can be proven that a formula η is satisfiable iff it is satisfiable in a Hintikka structure for the extended closure $ecl(\eta)$ of that formula.[37] A finite set of BML formulae Γ is satisfiable in a Hintikka structure for $ecl(\Gamma)$. The tableau procedure for a given input formula attempts to construct a graph representing sufficiently many possible Hintikka structures for the input formula.

Soundness of the tableau procedure means that if the input formula is satisfiable, then the final tableau is open. Completeness of the tableau procedure means that if the input formula is not satisfiable, then the final tableau is closed.[38] By contradiction, completeness means that if the final tableau is open, the input formula is satisfiable. The tableau procedure for testing satisfiability can be modified to perform (local) model-checking: Given a finite ITS $(\mathcal{T}, s)$ and a formula η, the model-checking tableau must decide whether $\mathcal{T}, s \models \eta$.

---

[37]Demri *et al.* (2016, p. 487).
[38]Demri *et al.* (2016, p. 496).

3.2.3.   *Basics of tableau construction in linear temporal logic*[39]

Basic logical constants and connectives in LTL are only $\top, \neg, \wedge$ as Boolean connectives and X, G, U as temporal operators. The other Boolean symbols, $\bot, \vee, \rightarrow$, and temporal operators, F, R, are defined in the usual way. In temporal logic, eventualities must be considered, such as something that will happen eventually in the future but without specifying exactly when. In LTL, eventualities are formulae of the type $\varphi \cup \psi$ (especially, $F\varphi$ and $\neg G\varphi$). The conjunctive, disjunctive, and successor formulae in LTL and their components are given in Fig. 3.4.

The tableau construction for LTL is analogous to that for BML, with the extension of the rules to include decompensation rules for the temporal operators in LTL (Fig. 3.5).

Conjunctive		Disjunctive		Successor	
Formula	Components	Formula	Components	Formula	Components
$\neg\neg\varphi$	φ, φ	$\varphi \vee \psi$	φ, ψ	$X\varphi$	φ
$\varphi \wedge \psi$	φ, ψ	$\varphi \rightarrow \psi$	$\neg\varphi, \psi$	$\neg X\varphi$	$\neg\varphi$
$G\varphi$	$\varphi, XG\varphi$	$\neg G\varphi$	$\neg\varphi, X\neg G\varphi$		
$\neg(\varphi U\psi)$	$\neg\psi, \neg\varphi \vee \neg X(\varphi U\psi)$	$\varphi U\psi$	$\psi, \varphi \wedge X(\varphi U\psi)$		

Fig. 3.4.   Types and components of formulae in LTL.[40]

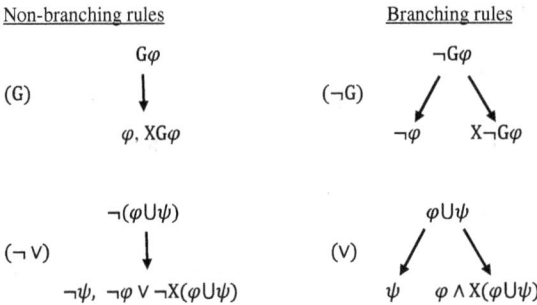

Fig. 3.5.   Branching and non-branching rules for the temporal operators in LTL.[41]

---

[39]Wolper (1983, 1985).
[40]Demri *et al.* (2016, p. 499).
[41]Demri *et al.* (2016, p. 499).

An example of an LTL formula is $\eta = (p \cup q) \wedge \mathrm{G}r$. The extended closure of η is $ecl(\eta) = \{\eta, p \cup q, \mathrm{G}r, p \wedge \mathrm{X}(p \cup q), q, r, \mathrm{XG}r, p, \mathrm{X}(p \cup q)\}$. In this case, there are two full expansions:

$$\Delta_1 = \{\eta, p \cup q, \mathrm{G}r, q, r, \mathrm{XG}r\},$$

$$\Delta_2 = \{\eta, p \cup q, \mathrm{G}r, \wedge \mathrm{X}(p \cup q), p, \mathrm{X}(p \cup q), r, \mathrm{XG}r\}.^{42}$$

For a given input formula $\eta \in$ LTL, the tableau procedure is used to test the satisfiability of η. When the procedure reaches a stabilization state, the tableau at that state is called the final tableau for η. The tableau procedure returns "not satisfiable" if the final tableau is closed, otherwise it returns "satisfiable." Soundness of the tableau construction means that if the input formula η is satisfiable, the final tableau is open. Completeness of the tableau for LTL means that for any formula $\eta \in$ LTL, if the final tableau is open, then the formula is satisfiable.[43]

The tableau construction for LTL can be adapted to TLR and CTL.[44] The syntax of TLR contains $\top, \neg, \wedge, \mathrm{EX}, \mathrm{AX}$, and EF as connectives, which are extended in CTL by EG, EU, and AU. The operators $\vee, \rightarrow, \mathrm{AG}, \mathrm{AF}, \mathrm{ER}$, and AR are assumed to be definable. The only eventualities in TLR are of type $\mathrm{EF}\varphi$, but there are two types of eventualities in CTL:

— existential eventualities: $\mathrm{EF}\varphi$ and $\mathrm{E}(\varphi \cup \psi)$,
— universal eventualities: $\neg\mathrm{EG}\varphi$ and $\mathrm{A}(\varphi \cup \psi)$.

The conjunctive, disjunctive, and successor formulae in CTL and their components are given in Fig. 3.6.

Decomposition rules for the temporal operators in CTL formulae are presented in Fig. 3.7.

The expansion procedure of a CTL formula by the application of the tableau rules in Fig. 3.7 is illustrated in the tree diagram of Fig. 3.8. To simplify notations, the CTL input formula is replaced

---

[42]Demri *et al.* (2016, p. 500).
[43]Demri *et al.* (2016, p. 510).
[44]Ben-Ari *et al.* (1981); Pratt (1979, 1980).

Conjunctive		Disjunctive	
Formula	Components	Formula	Components
$\neg\neg\varphi$	φ, φ	$\neg EF(\varphi \wedge \psi)$	$\neg\varphi, \neg\psi$
$\varphi \wedge \psi$	φ, ψ	$EF\varphi$	$\varphi, EXEF\varphi$
$\neg EF\varphi$	$\neg\varphi, \neg EXEF\varphi$	$\neg EG\varphi$	$\neg\varphi, \neg EXEG\varphi$
$EG\varphi$	$\varphi, EXEG\varphi$	$E(\varphi U\psi)$	$\psi, \varphi \wedge EXE(\varphi U\psi)$
$\neg E(\varphi U\psi)$	$\neg\psi, \neg\varphi \vee \neg EXE(\varphi U\psi)$	$A(\varphi U\psi)$	$\psi, \varphi \wedge AXA(\varphi U\psi)$
$\neg A(\varphi U\psi)$	$\neg\psi, \neg\varphi \vee \neg AXA(\varphi U\psi)$		

Successor	
Formula	Component
$EX\varphi$ (existential)	φ
$AX\varphi$ (universal)	φ
$\neg AX\varphi$ (existential)	$\neg\varphi$
$\neg EX\varphi$ (universal)	$\neg\varphi$

Fig. 3.6.   Types and components of formulae in CTL.

by its set of conjuncts:

$$\Gamma = \{EF\neg p, A(p \cup q), \neg p \rightarrow EG\neg q\} \text{ with } \varphi := EF\neg p, \psi := A(p \cup q),$$
and $\chi := EG\neg q$.

The extended closure of Γ contains the following formulae:

$$ecl(\Gamma) = \Gamma \cup \{\neg p, EX\varphi, q, p \wedge AX\psi, p, AX\psi, \chi, \neg q, EX\chi\}.$$

In the expansion tree of Fig. 3.7, the non-branching steps are represented in single nodes. Some decompositions are abbreviated as

$$\Psi_1 := \{p \wedge AX\psi, p, AX\psi\},$$
$$\Psi_2 := \{\chi, EX\chi, p, \neg q\}.$$

Non-branching rules   Branching rules

$$EG\varphi$$
(EG) $\downarrow$

$\varphi, EXEG\varphi$

$$\neg EG\varphi$$
$(\neg EG)$ $\swarrow$ $\searrow$

$\neg\varphi$ $EX\neg EG\varphi$

$$\neg EF\varphi$$
$(\neg EF)$ $\downarrow$

$\neg\varphi, \neg EXEF\varphi$

$$\neg EF\varphi$$
(EF) $\swarrow$ $\searrow$

$\neg\varphi$ $EXEF\varphi$

$$\neg E(\varphi\cup\psi)$$
$(\neg EU)$ $\downarrow$

$\neg\psi, \neg\varphi \vee \neg EXE(\varphi\cup\psi)$

$$E(\varphi\cup\psi)$$
(EU) $\swarrow$ $\searrow$

ψ $\varphi \wedge EXE(\varphi\cup\psi)$

$$\neg A(\varphi\cup\psi)$$
$(\neg AU)$ $\downarrow$

$\neg\psi, \neg\varphi \vee \neg AXA(\varphi\cup\psi)$

$$A(\varphi\cup\psi)$$
(AU) $\swarrow$ $\searrow$

ψ $\varphi \wedge AXA(\varphi\cup\psi)$

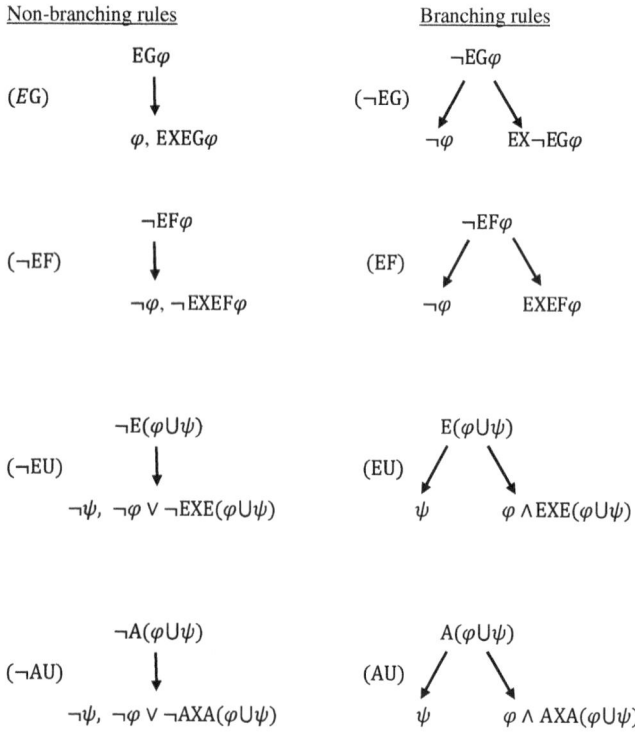

Fig. 3.7. Branching and non-branching rules for the temporal operators in CTL.[45]

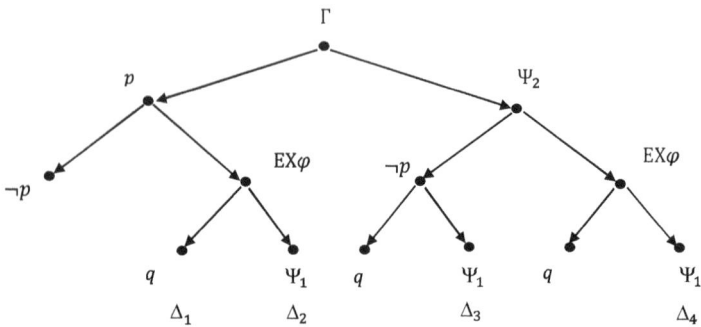

Fig. 3.8. Full expansion tree of a CTL formula with tableau procedure.[46]

---

[45]Demri *et al.* (2016, p. 516).
[46]Demri *et al.* (2016, p. 517).

Fully expanded sets of formulae are denoted by

$$\Delta_1 := \Gamma \cup \{EX\varphi, p, q\},$$

$$\Delta_2 := \Gamma \cup \Psi_1 \cup \{EX\varphi, p\},$$

$$\Delta_3 := \Gamma \cup \Psi_2 \cup \{EX\varphi, q\},$$

$$\Delta_4 := \Gamma \cup \Psi_1 \cup \Psi_2 \cup \{EX\varphi\}.$$

For a given input formula $\eta \in$ CTL, the tableau procedure can again be used to test the satisfiability of η. When the procedure reaches a stabilization state, the tableau at that state is called the final tableau for η. The tableau procedure returns "not satisfiable" if the final tableau is closed, otherwise it returns "satisfiable." Soundness of the tableau construction means again that if the input formula η is satisfiable, the final tableau is open. Completeness of the tableau for CTL means that for any formula $\eta \in$ CTL, if the final tableau is open, then the formula is satisfiable.[47] For full CTL* temporal logic, tableaux procedures were also suggested for generalized tree semantics.[48]

3.3. Automata-Based Calculus and Temporal Logic

A decision procedure in temporal logic can also be realized by automata. A decision procedure with automata consists of two steps. In the first step, an effective translation from logical formulae to certain automata must be found. In the second step, a corresponding decision problem of automata must be identified. In LTL, for example, an LTL formula describes a set of computation paths. From a language-theoretic point of view, computation paths are infinite words over an alphabet obtained as a power set of a finite set of atomic propositions. In this sense, we seek for transformations of LTL formulae into automata that recognizes languages over such words by visiting accepting states infinitely often. The language of the automaton $\mathcal{A}_\varphi$ constructed from the temporal formula φ recognizes exactly

---

[47]Demri *et al.* (2016, p. 535).
[48]Reynolds (2007, 2009, 2011, 2013).

those computation paths that are models of φ. Then, the satisfiability of φ reduces to the nonemptiness of the language of $\mathcal{A}_\varphi$ (i.e., the existence of a model for φ). The two steps of the effective translation from φ to $\mathcal{A}_\varphi$ and an algorithm for checking $\mathcal{A}_\varphi$ for nonemptiness deliver a decision procedure for the satisfiability of the temporal logic.

In Fig. 3.1, a tableau procedure for LTL formula $\varphi := \mathrm{GF}p \wedge \mathrm{GF}\neg p$ was discussed. Intuitively, φ means that p must hold infinitely often but also that p must not hold infinitely often. The satisfiability problem of φ must first be transferred to an appropriate automaton. The automaton in Fig. 3.9 accepts exactly those computation paths which can be considered infinite words over the alphabet $\{\emptyset, \{p\}\}$, which satisfy the condition expressed by φ. It can be verified that the trace $\{p\}\emptyset\emptyset\{p\}\emptyset\cdots$ is accepted by this automaton.[49]

In short, the automata-based procedure means the reduction of logical problems to automata-based decision problems in order to use the advantages of decision procedures in automata theory: The existence of a model for formula φ (satisfiability of φ) is reduced to checking the existence of a word, resp. tree, accepted by an automaton $\mathcal{A}_\varphi$ (nonemptiness of $\mathcal{A}_\varphi$). The truth of φ in all models (validity of φ) is reduced to checking whether automaton $\mathcal{A}_\varphi$ accepts all words, resp. trees (universality of $\mathcal{A}_\varphi$). Entailment of one formula φ by another formula ψ is reduced to checking whether all what is accepted by automaton $\mathcal{A}_\varphi$ is also accepted by the automaton $\mathcal{A}_\psi$.

The reduction of logical decision problems to automata-based decision problems was first realized by the so-called Büchi automata

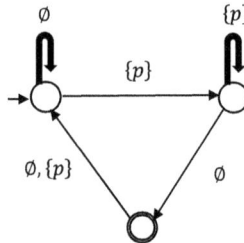

Fig. 3.9.   Automaton associated with the LTL formula $\mathrm{GF}p \wedge \mathrm{GF}\neg p$.

---

[49]Demri *et al.* (2016, pp. 470–471).

for the monadic second-order logic (MSO) over $(\mathbb{N}, <)$.[50] LTL formulae can be translated into alternating Büchi automata. CTL formulae can be translated into nondeterministic Büchi automata.[51]

3.3.1. *What are Büchi automata?*

A Büchi automaton is a finite-state automaton accepting infinite ω-words (i.e., ordinal number $\omega = \{0, 1, 2, \ldots\}$) instead of finite words. Formally, a Büchi automaton is defined as a tuple $\mathcal{A} = (\Sigma, Q, Q_0, \delta, F)$ with a finite alphabet Σ, a finite set Q of states, the set $Q_0 \subseteq Q$ of initial states, the transition relation δ as a subset of $Q \times \Sigma \times Q$, and a set $F \subseteq Q$ of accepting states. For $q \in Q$ and $a \in \Sigma$, the set of states q' with $(q, a, q') \in \delta$ is denoted by $\delta(q, a)$. Automaton $\mathcal{A}$ is deterministic iff $card(Q_0) = 1$ and $card(\delta(q, a)) \leq 1$ for all $q \in Q$ and $a \in \Sigma$.

A run ρ of the Büchi automaton $\mathcal{A}$ is a sequence $q_0 \xrightarrow{a_0} q_1 \xrightarrow{a_1} q_2 \cdots$, with $q_0 \in Q_0$ and $(q_i, a_i, q_{i+1}) \in \delta$ for every $i \geq 0$. The label of the run ρ is the word $w = a_0 a_1 \cdots \Sigma^\omega$. For a run ρ, the infimum is defined by

$$\inf(\rho) := \{q \in Q \mid \text{for all } i \in \mathbb{N} \text{ there exists } j > i \text{ such that } q = q_j\}.$$

The run ρ is accepting if some state of F is repeated infinitely often in ρ, i.e., $\inf(\rho) \cap F \neq \emptyset$.

Büchi acceptance condition: The automaton $\mathcal{A} = (\Sigma, Q, Q_0, \delta, F)$ accepts the language $L(\mathcal{A})$ which consists of all ω-words $w \in \Sigma^\omega$ for which there exists an accepting run of $\mathcal{A}$ with label w.

A language $L \subseteq \Sigma^\omega$ is Büchi-recognizable whenever there is a Büchi automaton $\mathcal{A}$ such that $L(\mathcal{A}) = L$. Figure 3.10 illustrates a Büchi automaton $\mathcal{A}$ with $L(\mathcal{A}) = \{w \in \{a, b\}^\omega \mid \text{the letter a occurs infinitely often}\}$. The accepting states are doubly circled. $\mathcal{A}$ accepts those words over $\{a, b\}$ which have infinitely many a's.

The size of a Büchi automaton $\mathcal{A}$ is defined by $|\mathcal{A}| := card(\Sigma) + card(Q_0) + card(\delta) + card(F)$. The decision algorithm for solving the nonemptiness problem $L(\mathcal{A}) \neq \emptyset$ can be realized by reachability checks.

---

[50] Büchi (1962).
[51] D'Souza and Shankar (2012).

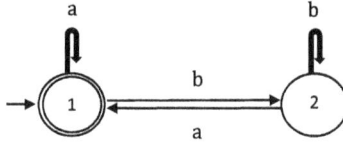

Fig. 3.10.   Büchi automaton accepting the words over $\{a, b\}$ which have infinitely many a's.[52]

3.3.2.   *Büchi automata for linear temporal logic*[53]

The automata-based approach for LTL represents the set $\text{Mod}(\varphi)$ of models of a formula φ by ω-words w for which there exists an accepting run of a Büchi automaton $\mathcal{A}$ with label w. Satisfiability of an LTL formula φ is equivalent to the nonemptiness of the set $\text{Mod}(\varphi)$ of the models of φ. Thus, satisfiability of an LTL formula is equivalent to the nonemptiness of a Büchi automaton.

The set of linear models of a formula φ over the the set Π of atomic propositions is denoted by $\text{Mod}_\Pi(\varphi)$. It can be proven that, for a given LTL formula φ, a Büchi automaton $\mathcal{A}_\varphi$ can be effectively constructed such that $\text{Mod}_\Pi(\varphi) = L(\mathcal{A}_\varphi)$, or in short, $\varphi \approx \mathcal{A}_\varphi$ for $\Pi = \text{PROP}(\varphi)$.

In the recursive construction of $\mathcal{A}_\varphi$, one has to first consider the Boolean connectives. It can be proven that Büchi-recognizable sets are effectively closed under union, intersection, and complementation. For LTL formulae φ_1 and φ_2, a finite set Π of atomic propositions with $\text{PROP}(\varphi_1) \cup \text{PROP}(\varphi_1) \subseteq \Pi$, Büchi automata $\mathcal{A}_1$ and $\mathcal{A}_2$ with $\varphi_1 \approx_\Sigma \mathcal{A}_1$ and $\varphi_2 \approx_\Sigma \mathcal{A}_2$, and a Büchi automaton $\mathcal{B}$ over alphabet Σ, it holds that

(i) if $L(\mathcal{B}) = L(\mathcal{A}_1) \cup L(\mathcal{A}_2)$, then $\varphi_1 \vee \varphi_2 \approx_\Sigma \mathcal{B}$;

(ii) if $L(\mathcal{B}) = L(\mathcal{A}_1) \cap L(\mathcal{A}_2)$, then $\varphi_1 \wedge \varphi_2 \approx_\Sigma \mathcal{B}$;

(iii) if $L(\mathcal{B}) = \Sigma^\omega \setminus L(\mathcal{A}_1)$, then $\neg\varphi_1 \approx_\Sigma \mathcal{B}$.

The cases for the temporal operators X, F, and U must be treated separately. For an LTL formula φ, a finite set Π of

---

[52]Demri *et al.* (2016, p. 547).
[53]Vardi and Wolper (1986, 1994).

atomic propositions with $\mathrm{PROP}(\varphi) \subseteq \Pi$ and a Büchi automaton $\mathcal{A} = (\Sigma, Q, Q_0, \delta, F)$ with $\varphi \approx_\Sigma \mathcal{A}$ and $\Sigma = \mathcal{P}(\Pi)$, it holds that

(i) $\mathrm{X}_\varphi \approx_\Sigma \mathcal{A}'$ for a Büchi automaton $\mathcal{A}' := (\Sigma, Q \uplus \{q_{new}\}, \{q_{new}\}, \delta', F)$ obtained from $\mathcal{A}$ by adding a new state q_{new} with $\delta' := \delta \uplus \{q_{new} \overset{a}{\to} q_0 | a \in \Sigma, q_0 \in Q_0\}$;

(ii) $\mathrm{F}_\varphi \approx_\Sigma \mathcal{A}'$ for a Büchi automaton $\mathcal{A}' := (\Sigma, Q \uplus \{q_{new}\}, Q_0 \cup \{q_{new}\}, \delta', F)$ obtained from $\mathcal{A}$ by adding a new state q_{new} with $\delta' := \delta \uplus \{q_{new} \overset{a}{\to} q_0, q_{new} \overset{a}{\to} q_0 | a \in \Sigma, q_0 \in Q_0\}$.[54]

For LTL formulae φ_1 and φ_2, a finite set Π of atomic propositions with $\mathrm{PROP}(\varphi_1) \cup \mathrm{PROP}(\varphi_1) \subseteq \Pi$, and Büchi automata $\mathcal{A}_1$ and $\mathcal{A}_2$ with $\varphi_1 \approx_\Sigma \mathcal{A}_1$ and $\varphi_2 \approx_\Sigma \mathcal{A}_2$ with $\Sigma = \mathcal{P}(\Pi)$, there exists a Büchi automaton $\mathcal{B}$ over alphabet Σ with $\varphi_1 \cup \varphi_2 \approx_\Sigma \mathcal{B}$.

The advantage of inductive construction of an automaton $\mathcal{A}_\varphi$ from a formula φ is that we can apply the standard procedures of automata theory. $\mathcal{A}_\varphi$ accepts exactly the linear models from $\mathrm{Mod}(\varphi)$. In some cases, a disadvantage is the size of $\mathcal{A}_\varphi$ in the size of φ. The number of states can be written as an exponential tower $2^{2^{\cdot^{\cdot^{2^{p(|\varphi|)}}}}}$ with linear height and a polynomial φ.

3.3.3. Büchi tree automata for branching-time logic CTL

Until now, we considered translations from LTL formulae to Büchi automata accepting the models of the formulae. The idea is that LTL models are ω-words and Büchi automata accept these models. In branching-time logic, it is well known that CTL formulae are satisfiable iff they are satisfiable in infinite tree-like models with branching factor bounded by the size of the formulae.

An infinite Σ, k-tree Tr is a mapping Tr $:[1,k]^* \to \Sigma$. A branch Br starts at a node with a maximal (infinite) sequence such that for two consecutive nodes, one is the predecessor of the other.[55] A Büchi tree automaton $\mathcal{A}$ for Σ, k-trees is a tuple $\mathcal{A} = (\Sigma, Q, Q_0, \delta, F)$ with

---

[54]Demri *et al.* (2016, pp. 555–556).

[55]Tree automata, in general, were studied by Comon-Lundh *et al.* (2005).

a finite (nonempty) set Q of states, set $Q_0 \subseteq Q$ of initial states, the transition relation $\delta \subseteq Q \times \Sigma \times Q^k$, and set $F \subseteq Q$ of accepting states.

A run ρ of the Büchi tree automaton $\mathcal{A} = (\Sigma, Q, Q_0, \delta, F)$ on a Σ, k-tree Tr is a Q, k-tree with $\rho(\varepsilon) \in Q_0$ and $(\rho(u), \text{Tr}(u), \rho(u \cdot 1), \ldots, \rho(u \cdot k)) \in \delta$ for every $u \in [1, k]^*$. A run is accepting iff for every branch in Tr, there is a state in F that occurs infinitely often. The automaton $\mathcal{A}$ accepts the language $\text{L}(\mathcal{A})$ of the infinite Σ, k-tree Tr for which there is an accepting run of $\mathcal{A}$ on Tr.

In Fig. 3.11, an automaton $\mathcal{A} = (\Sigma, \{q_1, q_2\}, \{q_1\}, \delta, \{q_1\})$ is considered with alphabet $\Sigma = \{\text{a}, \text{b}\}$, $\delta = \{(q_1, \text{a}, q_1, q_1), (q_2, \text{a}, q_1, q_1), (q_1, \text{a}, q_2, q_2), (q_2, \text{a}, q_2, q_2)\}$, $k = 2$, and the set L of infinite Σ, k-trees such that on every branch, the letter a occurs infinitely often.[56] $\text{L}(\mathcal{A})$ is equal to language L.

The nonemptiness problem for Büchi tree automata is decidable in polynomial time.

When a nondeterministic automaton is in a state q and reads a letter a, then the transition function associates a set of possible states to continue the run. The universal mode as a dual of the existential mode in nondeterministic automata requires that all the runs from a location in the set of possible states should lead to acceptance. In alternating automata, both existential and universal modes are allowed, and the transitions are defined by positive Boolean formulae on the set of states. The Boolean formulae are defined with atomic formulae connected with disjunction $\vee$ and conjunction $\wedge$ only but

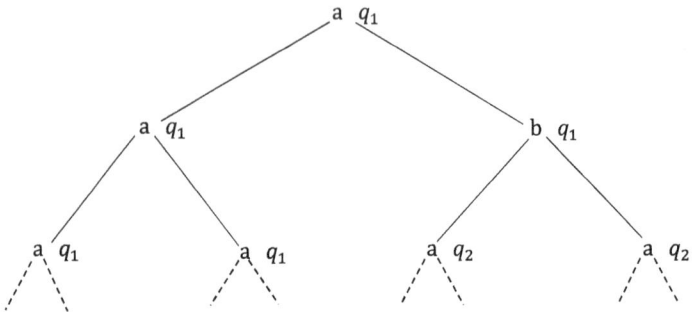

Fig. 3.11. $\Sigma, 2$-tree with its accepting run in an automaton $\mathcal{A}$.

---

[56]Demri *et al.* (2016, pp. 599–600, 602).

without negation $\neg$. Thus, a transition formula $q_1 \vee q_2$ says that at the next stage, there is a run starting with q_1 or q_2. A transition formula $q_1 \wedge q_2$ says that at the next stage, there are at least two runs, one starting with state q_1 and the other starting with q_2. Runs are words of nondeterministic automata which are replaced by trees for alternating automata because of the universal mode.

Formally, in a nondeterministic Büchi automaton, the transition relation $\delta \subseteq Q \times \Sigma \times Q$ is understood as a transition function of type $Q \times \Sigma \to \mathcal{P}(Q)$. Alternating automata use transition functions of type $Q \times \Sigma \to \mathbb{B}^+(Q)$ with the set $\mathbb{B}^+(Q)$ of positive Boolean formulae.[57] An alternating automaton is:

(i) deterministic iff the image of transition function δ is restricted to single states;

(ii) nondeterministic iff the image of transition function δ is restricted to disjunctions of single states or to $\bot$;

(iii) universal iff the image of transition function δ is restricted to conjunctions of single states or to $\top$.

For a translation from CTL formulae to automata which recognize branching trees, the class of alternating Büchi tree automata is introduced. In alternating automata, both existential and universal modes are possible, and the transition function is defined by positive Boolean formulae. In alternating Büchi tree automata, the atomic formulae are not only related to states but to states with directions. The directions are defined by elements in the interval $[1, k]$.

Formally, an alternating Büchi tree automaton $\mathcal{A}$ for Σ, k-trees is a tuple $\mathcal{A} = (\Sigma, Q, q_0, \delta, F)$ with a finite (nonempty) set Q of states, an initial state $q_0 \in Q$, and a transition function $\delta : Q \times \Sigma \to \mathbb{B}^+([1, k] \times Q)$ with the set $\mathbb{B}^+([1, k] \times Q)$ of positive Boolean formulae built over $[1, k] \times Q$.

In that way, transitions can lead to several copies of the automaton. For example, consider an alternating Büchi tree automaton $\mathcal{A}$ with $\delta(q, a) = (1, q_1) \wedge (1, q_2) \wedge (2, q_3) \wedge (2, q_4)$. When $\mathcal{A}$ is in state q and reads the node u of a Σ, k-tree Tr with Tr$(u) = a$, then $\mathcal{A}$ has to satisfy the positive Boolean formula $\delta(q, a)$ with several copies of

---

[57] A translation of temporal logics into alternating automata is given by Muller *et al.* (1988).

itself. One copy moves to the successor node $u \cdot 1$ with state q_1 and another one with state q_2. Another copy moves to the successor node $u \cdot 2$ with state q_3 and another one with state q_4.

A CTL formula φ can be translated into an alternating Büchi tree automaton $\mathcal{A}_\varphi$ such that φ is satisfiable if and only if $L(\mathcal{A}_\varphi) \neq \emptyset$.[58] For a formal inductive proof of the structure of φ, the CTL formula φ is assumed in negation normal form with Boolean connectives $\wedge$ and $\vee$, temporal operators EX, AX, EU, ERAX, AU, and AR, and negation only in front of atomic propositions. Furthermore, it can be proven that $L(\mathcal{A}_\varphi)$ accepts exactly the Σ, k-trees which satisfy φ. A formula φ is CTL satisfiable if and only if $L(\mathcal{A}_\varphi) \neq \emptyset$.

3.4.   Game-Based Calculus and Temporal Logic

After Gentzen's sequent calculus and tableaux- and automata-based procedures, game-based procedures deliver another tool kit for solving decision problems of temporal logic. Furthermore, from a philosophical point of view, model-checking games are also of great value to understand the meaning (semantics) of logical connectives, quantifiers, and modal and temporal operators. A game is played between two players on an underlying ITS and formula. The two players decompose the formula and check the states in the ITS in order to decide whether it holds. The simplest building block of a formula is a literal, i.e., atomic propositions and their negations. The truth value of a literal is obvious. Therefore, if the game has reached a position with a literal, then it ends, and the winner is determined.

3.4.1.   *Game-based calculus of basis modal logic*

At first, we consider BML formula in negation normal form with literals, logical connectives $\wedge, \vee$ and modal operators EX, AX. The existential operators in BML are disjunction and existential modalities. The universal operators in BML are conjunction and universal modalities. A disjunction holds in a state if one of the two disjuncts holds there. Thus, if a proponent of a disjunction is attacked by an opponent, then the proponent must defend the disjunction

---

[58]Demri *et al.* (2016, p. 609).

with some disjunct. The satisfaction of the disjunction is witnessed by some disjunct. A conjunction holds in a state if all the two disjuncts hold there. Thus, if a proponent of a conjunction is attacked by an opponent with a request to defend one of the conjuncts, then the proponent must be able to defend both conjuncts. The satisfaction of a conjunction is witnessed by all conjuncts, but it is refuted by some conjunct. The disjunction is refuted by all disjuncts.

In the case of modal operators, the satisfaction of $EX\,\psi$ at some state s is witnessed by some successor of s that satisfies ψ. $EX\,\psi$ can only be refuted by those successors which do not satisfy ψ. In a model-checking game, one player aims at the satisfaction of a given formula by a given ITS. The other player attempts to prove the opposite. Therefore, the players are sometimes called "verifier" (V) and "refuter" (R). A game can also be considered as a dialogue between a "proponent" (P) and an "opponent" (O), defending and attacking propositions according to the dialogue rules.

Formally, for a rooted $ITS(\mathcal{T}, s_0)$ with $\mathcal{T} = (S, \rightarrow, L)$ and a BML formula φ in negation normal form, the model-checking game $\mathcal{G}^{\mathcal{T}}(s_0, \varphi)$ is defined as a two-player game in a game environment (V, Own, E) with $V = S \times sub(\varphi)$ and $Own(s, \psi) = 0$ iff ψ is of the form $\psi_1 \vee \psi_2$ or $EX\,\psi$.

The development of the game is represented in a game tree. The nodes are pairs of states in the transition system ITS and subformulae. The satisfaction relation between a state s and a formula ψ, which must be checked at any node, is denoted by $s \vdash \psi$. The rules of the BML model-checking games are given in in Fig. 3.12.[59] The terminology underlines the similarity of the game-based calculus with Gentzen's sequent calculus. At a node with disjunctions, player

$$(\vee) \quad \frac{s \vdash \psi_1 \vee \psi_2}{s \vdash \psi_i} \;\; \mathbf{V} \qquad (\wedge) \quad \frac{s \vdash \psi_1 \wedge \psi_2}{s \vdash \psi_i} \;\; \mathbf{R}$$

$$(EX) \quad \frac{s \vdash EX\psi}{t \vdash \psi} \;\; \mathbf{V}, s \rightarrow t \qquad (EX) \quad \frac{s \vdash AX\psi}{t \vdash \psi} \;\; \mathbf{R}, s \rightarrow t$$

Fig. 3.12.   The BML model-checking game rules.

---

[59]Demri *et al.* (2016, p. 120).

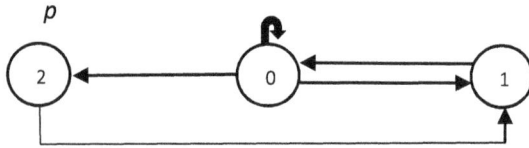

Fig. 3.13.   Graph of ITS for formula $\mathrm{AX}\,(p \vee \mathrm{EX}\,p)$.

V (verifier) chooses one of the disjuncts. At a node with conjunctions, player **R** (refuter) chooses one of the conjuncts. In the case of a modal operator, the players must choose a successor state of the current one. Then, the successor state and the formula following the modal operator must be checked, depending on whether it is the existential or the universal operator.

The winning conditions in BML model-checking games are defined for both players **V** and **R**. A play in the model-checking game $\mathcal{G}^{\mathcal{T}}(s_0, \varphi)$ is denoted by a sequence of nodes, $\pi = \pi_0, \ldots, \pi_n$. The final node π_n cannot have a successor node. According to the game rules, π_n must represent one of four possible configurations to define a winner of the play:

Player **V** wins the play π iff

(i) there is a state t and an atomic q such that $\pi_n = t \vdash q$ and $q \in L(t)$, or $\pi_n = t \vdash \neg q$ and $q \notin L(t)$;

(ii) there is a state t and a formula ψ such that $\pi_n = t \vdash \mathrm{AX}\,\psi$ and t has no successor in $\mathcal{T}$.

Player **R** wins the play π iff

(i) there is a state t and an atomic q such that $\pi_n = t \vdash q$ and $q \notin L(t)$, or $\pi_n = t \vdash \neg q$ and $q \in L(t)$;

(ii) there is a state t and a formula ψ such that $\pi_n = t \vdash \mathrm{EX}\,\psi$ and t has no successor in $\mathcal{T}$.

In order to illustrate winning strategies in model-checking games, the BML formula $\varphi = \mathrm{AX}(p \vee \mathrm{EX}\,p)$ is considered with respect to the (interpreted) transition system $\mathcal{T}$ with the graph shown in Fig. 3.13.

The game tree of the model-checking game $\mathcal{G}^{\mathcal{T}}(0, \varphi)$ with $\mathcal{T}$ is represented in Fig. 3.14, which also considers the games $\mathcal{G}^{\mathcal{T}}(1, \varphi)$ and $\mathcal{G}^{\mathcal{T}}(2, \varphi)$. The task is to check at which state, 0, 1, or 2, the LTL formula φ is satisfied. In Fig. 3.14, the ending (winning) nodes of player **R** are denoted by rounded rectangles and those of player

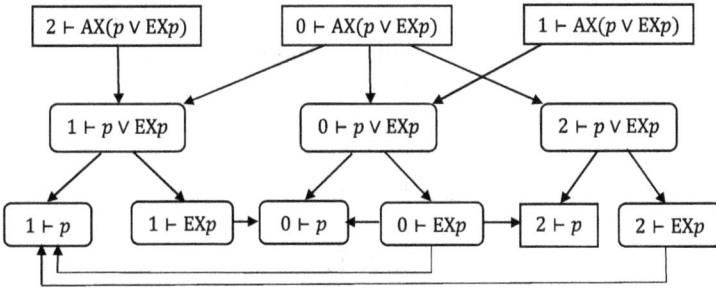

Fig. 3.14. BML model-checking game.

V by rectangles.[60] Obviously, **R** has a winning strategy which starts from the sequent $0 \vdash AX(p \vee EXp)$. At first, the strategy moves to $1 \vdash p \vee EX p$. Independent of the next step or the next two steps of player **V**, the strategy arrives at $0 \vdash p$ or $1 \vdash p$. In any case, **R** wins the game according to the mentioned winning conditions.

It can be proven that, for a model-checking game $\mathcal{G}^{\mathcal{T}}(s_0, \varphi)$ for an (interpreted) transition system $\mathcal{T}$ with root s_0 and a BML formula φ, it holds that $\mathcal{T}, s_0 \vdash \varphi$ iff player **V** has a winning strategy for the game starting with $s_0 \vdash \varphi$.

3.4.2. Game-based calculus and certification of computer programs

As in a tableaux-based and automata-based interpretation, the game-based approach is important for the decision problems of temporal logic. Model-checking and satisfiability-checking games can be used to solve the model-checking and satisfiability-checking problems for temporal logic. A model-checking game clarifies whether a given state of an ITS satisfies a temporal formula by proving the existence of a winning strategy for the proponent of the formula. A satisfiability-checking game clarifies whether a given formula is satisfiable by checking whether the proponent of the formula has a winning strategy.

The game-theoretic approach is helpful for program verification in computer science. The model-checking problem can be used to

---

[60]Demri *et al.* (2016, pp. 121–122).

prove whether a program is correct with respect to a property which is represented in a temporal logic. If the property is satisfied in the sense that model-checking returns true, then the program is correct in this case, and checks can be continued for further properties of the program. If model-checking returns false, then the cause of non-satisfaction must be clarified. In engineering and programming praxis, engineers and programmers try to find the causes of errors or bias in a dialogue with experts. Therefore, a game-based dialogue between a proponent and an opponent is an appropriate procedure to simulate the search for causes of failures. The game-based approach is useful to deliver explanations. Winning strategies justify certifications that a program specification is realizable. After satisfiability of temporal formulas in BML, it is a challenge to certify temporal formulas of LTL, CTL, and CTL*. Again, in these extended versions of temporal logic, winning strategies must be found for certification with satisfying models.

3.4.3. *Game-based solution to the model-checking problem for CTL**[61]

For model-checking games for the full branching-time logic CTL*, we must consider the structure of a CTL* formula connected with a total rooted transition system. Given a rooted transition system $\mathcal{T} = (S, \rightarrow, L, s_0)$, a CTL* formula ϑ in negation normal form consists of a path quantifier E or A, followed by a linear-time formula (LTL) over propositional literals or smaller blocks. The CTL* model-checking game is applied to a closure of formulae which are made up from the input formula. The Fischer–Ladner closure $fi(\varphi)$ of CTL* formula φ in negation normal form is defined as the smallest set that satisfies the following conditions[62]:

(i) If $\psi_1 \vee \psi_2 \in fi(\varphi)$ or $\psi_1 \wedge \psi_2 \in fi(\varphi)$, then $\{\psi_1, \psi_2\} \subseteq fi(\varphi)$

(ii) If $Q\psi \in fi(\varphi)$ for some $Q \in \{E, A, X\}$, then $\psi \in fi(\varphi)$.

(iii) If $\psi_1 \cup \psi_2 \in fi(\varphi)$, then $\{\psi_1, \psi_2, X(\psi_1 \cup \psi_2)\} \subseteq fi(\varphi)$.

(iv) If $\psi_1 R \psi_2 \in fi(\varphi)$, then $\{\psi_1, \psi_2, X(\psi_1 R \psi_2)\} \subseteq fi(\varphi)$.

---

[61]Lange and Stirling (2000, pp. 115–125).
[62]Fisher and Ladner (1979, pp. 194–211).

The model-checking game for the ITS $\mathcal{T} = (S, \to, L, s_0)$ and a CTL* formula ϑ in negation normal form is defined by the following conditions:

(i) With respect to Gentzen's notation, the configurations of the game are in the form of sequent $s \vdash \mathrm{E}(\Gamma)$ or $s \vdash \mathrm{A}(\Delta)$ with $s \in S$ and $\Gamma, \Delta \subseteq fi(\varphi)$. (Abbreviations are used such as $\mathrm{E}\,\psi$ instead of $\mathrm{E}(\{\psi\})$ and $\mathrm{A}(\psi, \Delta)$ instead of $\mathrm{A}(\{\psi\} \cup \Delta)$.)

(ii) The game starts with the sequent $s \vdash \mathrm{A}(\vartheta)$. (The A underlines that the formula component of every configuration begins with a path quantifier.)

(iii) The game rules are again represented in a Gentzen-style notation (Fig. 3.15).[63]

A formula set under an E-quantifier, resp. A-quantifier, must be interpreted as a conjunction, resp. disjunction. Therefore, a sequent $s \vdash \mathrm{E}\Gamma$ can be considered as **V**'s strategy to begin a path in s on which all formulae in Γ are true. A sequent $s \vdash \mathrm{A}\Delta$ must be considered as **R**'s obligation to begin a path in s in which none of the formulae in Δ is true.

$$(\mathrm{E}\wedge)\ \frac{s\vdash\mathrm{E}(\varphi\wedge\psi,\ \Gamma)}{s\vdash\mathrm{E}(\varphi,\psi,\Gamma)} \qquad (\mathrm{EU})\ \frac{s\vdash\mathrm{E}(\varphi\mathrm{U}\,\psi,\Gamma)}{s\vdash\mathrm{E}(\psi,\Gamma)\ \ s\vdash\mathrm{E}(\varphi,\mathrm{X}(\varphi\mathrm{U}\psi),\Gamma)}\ \mathbf{V}$$

$$(\mathrm{A}\vee)\ \frac{s\vdash\mathrm{A}(\varphi\vee\psi,\Delta)}{s\vdash\mathrm{A}(\varphi,\psi,\Delta)} \qquad (\mathrm{AU})\ \frac{s\vdash\mathrm{A}(\varphi\mathrm{U}\,\psi,\Delta)}{s\vdash\mathrm{A}(\psi,\varphi,\Delta)\ \ s\vdash\mathrm{A}(\varphi,\mathrm{X}(\varphi\mathrm{U}\psi),\Delta)}\ \mathbf{R}$$

$$(\mathrm{ALit})\ \frac{s\vdash\mathrm{A}(\ell,\Delta)}{s\vdash\mathrm{A}\Delta}\ \text{if}\ s\not\models\ell \qquad (\mathrm{ER})\ \frac{s\vdash\mathrm{E}(\varphi\mathrm{R}\,\psi,\Gamma)}{s\vdash\mathrm{E}(\psi,\varphi,\Gamma)\ \ s\vdash\mathrm{E}(\varphi,\mathrm{X}(\varphi\mathrm{R}\psi),\Gamma)}\ \mathbf{V}$$

$$(\mathrm{ELit})\ \frac{s\vdash\mathrm{E}(\ell,\Gamma)}{s\vdash\mathrm{E}\Gamma}\ \text{if}\ s\models\ell \qquad (\mathrm{AR})\ \frac{s\vdash\mathrm{A}(\varphi\mathrm{R}\,\psi,\Delta)}{s\vdash\mathrm{A}(\psi,\Delta)\ \ s\vdash\mathrm{A}(\varphi,\mathrm{X}(\varphi\mathrm{R}\psi),\Delta)}\ \mathbf{R}$$

$$(\mathrm{E}\vee)\ \frac{s\vdash\mathrm{E}(\varphi\vee\psi,\Gamma)}{s\vdash\mathrm{E}(\varphi,\Gamma)\ \ s\vdash\mathrm{E}(\psi,\Gamma)}\ \mathbf{V} \qquad (\mathrm{A}\wedge)\ \frac{s\vdash\mathrm{A}(\varphi\wedge\psi,\Delta)}{s\vdash\mathrm{A}(\varphi,\Delta)\ \ s\vdash\mathrm{A}(\psi,\Delta)}\ \mathbf{R}$$

$$(\mathrm{AQ})\ \frac{s\vdash\mathrm{A}(\mathrm{Q}\psi,\Delta)}{s\vdash\mathrm{A}\Delta\ \ s\vdash\mathrm{Q}\psi}\ \mathbf{V} \qquad (\mathrm{EQ})\ \frac{s\vdash\mathrm{E}(\mathrm{Q}\psi,\Gamma)}{s\vdash\mathrm{E}\Gamma\ \ s\vdash\mathrm{Q}\psi}\ \mathbf{R}$$

$$(\mathrm{AX})\ \frac{s\vdash\mathrm{A}(\mathrm{X}\varphi_1,...,\mathrm{X}\varphi_k)}{t\vdash\mathrm{A}(\varphi_1,...,\varphi_k)}\ \mathbf{R}\!:\!s\to t \qquad (\mathrm{EX})\ \frac{s\vdash\mathrm{E}(\mathrm{X}\varphi_1,...,\mathrm{X}\varphi_k)}{t\vdash\mathrm{E}(\varphi_1,...,\varphi_k)}\ \mathbf{V}\!:\!s\to t$$

Fig. 3.15.   Rules for the CTL* model-checking game.

---

[63]Lange and Stirling (2002, pp. 623–639), Demri *et al.* (2016, p. 667).

The winning conditions for finite plays are defined for a transition system $\mathcal{T}$ and a propositional literal ℓ:

V wins a finite play, which reaches a sequent of the form

(i) $s \vdash E\emptyset$ or
(ii) $s \vdash A(\ell, \Delta)$ such that $\mathcal{T}, s \models \ell$.

R wins a finite play, which reaches a sequent of the form

(i) $s \vdash A\emptyset$ or
(ii) $s \vdash E(\ell, \Gamma)$ such that $\mathcal{T}, s \not\models \ell$.

The winning conditions for final plays are related to configurations in which a sequent is true or false with appropriate interpretation. In these cases, the formulae cannot be decomposed further by the game rules.

An infinite play of a CTL* model-checking game consists of an infinite sequence of sequents $s_0 \vdash Q_0\Phi_0$, $s_1 \vdash Q_1\Phi_1, \ldots$. In an E-play, there is some $n \in \mathbb{N}$ with $Q_i = E$ for all $i \geq n$. In an A-play, there is some $n \in \mathbb{N}$ with $Q_i = A$ for all $i \geq n$. It can be proven that every play is either an E-play or an A-play.

Formulae of successive sequents can be related by a connection relation.[64] If a sequent $s' \vdash Q\Phi'$ is derived from sequent $s \vdash Q\Phi$ by a certain rule, then some $\varphi \in \Phi$ is said to be connected to $\psi \in \Phi'$ if

(i) the rule replaces φ by ψ inside the outer path quantifier (i.e., rules (E$\wedge$), (A$\vee$), (E$\vee$) and (A$\vee$) with conjunctions and disjunctions and rules (E$\cup$), (A$\cup$), (ER), and (AR) with $\cup$-formulae and R-formulae), or
(ii) the rule replaces φ by ψ, which becomes the new outer path quantifier (i.e., rules (AQ) and (EQ) with formula $Q\psi$), or
(iii) $\varphi = X\psi$ with rule (EX) or (AX), or
(iv) $\varphi = \psi$, and the rule operates on a formula different to φ.

A thread is an infinite sequence $\varphi_0, \varphi_1, \ldots$ in an infinite play $s_0 \vdash Q_0\Phi_0$, $s_1 \vdash Q_1\Phi_1, \ldots$ in a model-checking game for $\mathcal{T}, s_0$, and ϑ, with $\varphi_i \in \Phi_i$ and φ_i connected to φ_{i+1} for all $i \in \mathbb{N}$. A thread is called a μ-thread if there is some formula φ of the form $\psi_1 \cup \psi_2$ and $\varphi = \varphi_i$

---

[64]Demri *et al.* (2016, pp. 670–671).

for infinitely many i. A thread is called a ν-thread if there is some formula φ of the form $\psi_1 R \psi_2$. It can be proven (with König's lemma) that every infinite play has a thread.

The winning conditions for infinite plays are defined with μ-threads and ν-threads:

V wins an infinite play λ in the model-checking game for CTL* if

(i) λ is an E-play and contains no μ-thread, or
(ii) λ is an A-play and contains a ν-thread.

R wins an infinite play λ in the model-checking game for CTL* if

(i) λ is an E-play and contains a μ-thread, or
(ii) λ is an A-play and contains no ν-thread.

The model-checking problem for CTL* can be solved by model-checking games. Soundness is guaranteed because whenever **V** has a winning strategy in the model-checking game, then the underlying formula is satisfied in the transition system. Soundness can be proven because whenever the underlying formula is satisfied in the transition system, then **V** has a winning strategy in the model-checking game.

3.4.4. Game-based solution to the satisfiability problem for CTL*[65]

After the model-checking problem, the satisfiability problem should be solved by a game-theoretic characterization for the full branching-time logic CTL*. A configuration for the CTL* satisfiability game $\mathcal{G}(\vartheta)$ on a CTL* state formula ϑ in negation normal form is characterized by a set consisting of sets of formulae

$$E\Gamma_1, \ldots, E\Gamma_n, A\Delta_1, \ldots, A\Delta_m, \Lambda,$$

with sets $\Delta_i, \Gamma_j \subseteq fi(\vartheta)$ for $i = 1, \ldots, m$ and $j = 1, \ldots, n$ and set Λ of literals.

The winning conditions of satisfiability games for finite plays are defined by a formula ϑ in negation normal form.

---

[65]Friedmann *et al.* (2013, pp. 1–36).

V wins a play of configurations (sequents) $C_0, C_1, \ldots$ in satisfiability game $\mathcal{G}(\vartheta)$ if there is an $n \in \mathbb{N}$ such that C_n consists of literals $\ell_0, \ldots, \ell_k$ over atomic propositions without i, j such that $\ell_i = \ell_j$.

R wins a play of configurations (sequents) $C_0, C_1, \ldots$ in satisfiability game $\mathcal{G}(\vartheta)$ if there is an $n \in \mathbb{N}$ such that either

(i) there is some atomic proposition p with $\{p, \neg p\} \subseteq C_n$, or
(ii) $\mathrm{A}\emptyset \in C_n$.

With respect to the complexity of configurations in satisfiability games CTL*, there must be rules for every logical operator in any position of a configuration.[66] In Fig. 3.16, a set of the form Γ is interpreted conjunctively. A set of the form Δ is interpreted disjunctively. $Q\varphi$ represents a path-quantified formula with $\mathrm{E}\varphi$ or $\mathrm{A}\varphi$. $\mathrm{X}\Gamma$, resp. $\mathrm{X}\Delta$, means a set of formulae with each formula in Γ, resp. Δ, with an X-operator in front. In rules (X) and (X$^-$), Λ is assumed without a proposition and its negation because otherwise the configuration would be final and winning for **R**.

A block in a configuration is a literal or a formula set $\mathrm{E}\Gamma$ or $\mathrm{A}\Delta$. In a satisfiability game $\mathcal{G}(\vartheta)$, a connection relation can be defined between blocks B and B' of successive configurations C_i, C_{i+1} in a

$$\text{(E } \wedge) \quad \frac{\mathrm{E}(\varphi \wedge \psi, \ \Gamma), \Phi}{\mathrm{E}(\varphi, \psi, \Gamma), \Phi} \qquad \text{(EU)} \quad \frac{\mathrm{E}(\varphi \mathrm{U} \psi, \Gamma), \Phi}{\mathrm{E}(\psi, \Gamma), \Phi \quad \mathrm{E}(\varphi, \mathrm{X}(\varphi \mathrm{U} \psi), \Gamma), \Phi} \quad \mathbf{V}$$

$$\text{(A } \vee) \quad \frac{\mathrm{A}(\varphi \vee \psi, \Delta), \Phi}{\mathrm{A}(\varphi, \psi, \Delta), \Phi} \qquad \text{(AU)} \quad \frac{\mathrm{A}(\varphi \mathrm{U} \psi, \Delta), \Phi}{\mathrm{A}(\psi, \varphi, \Delta) \quad \mathrm{A}(\varphi, \mathrm{X}(\varphi \mathrm{U} \psi), \Delta), \Phi} \quad \mathbf{R}$$

$$\text{(ELit)} \quad \frac{\mathrm{E}(\ell, \Gamma), \Phi}{\mathrm{E}\Gamma, \Phi, \ell} \qquad \text{(ER)} \quad \frac{\mathrm{E}(\varphi \mathrm{R} \psi, \Gamma), \Phi}{\mathrm{E}(\psi, \varphi, \Gamma), \Phi \quad \mathrm{E}(\varphi, \mathrm{X}(\varphi \mathrm{R} \psi), \Gamma), \Phi} \quad \mathbf{V}$$

$$\text{(EQ)} \quad \frac{\mathrm{E}(\mathrm{Q}\psi, \Gamma), \Phi}{\mathrm{E}\Gamma, \mathrm{Q}\psi, \Phi} \qquad \text{(AR)} \quad \frac{\mathrm{A}(\varphi \mathrm{R} \psi, \Delta), \Phi}{\mathrm{A}(\psi, \Delta) \quad \mathrm{A}(\varphi, \mathrm{X}(\varphi \mathrm{R} \psi), \Delta), \Phi}$$

$$\text{(E } \vee) \quad \frac{\mathrm{E}(\varphi \vee \psi, \Gamma), \Phi}{\mathrm{E}(\varphi, \Gamma), \Phi \quad \mathrm{E}(\psi, \Gamma), \Phi} \quad \mathbf{V} \qquad \text{(A } \wedge) \quad \frac{\mathrm{A}(\varphi \wedge \psi, \Delta), \Phi}{\mathrm{A}(\varphi, \Delta), \ \mathrm{A}(\psi, \Delta), \Phi}$$

$$\text{(ALit)} \quad \frac{\mathrm{A}(\ell, \Delta), \Phi}{\mathrm{A}\Delta, \Phi \quad \Phi, \ell} \quad \mathbf{V} \qquad \text{(AQ)} \quad \frac{\mathrm{A}(\mathrm{Q}\varphi, \Delta), \Phi}{\mathrm{A}\Delta, \Phi \quad \mathrm{Q}\varphi, \Phi} \quad \mathbf{V}$$

$$\text{(X)} \quad \frac{\mathrm{EX}\Gamma_1, \ldots, \mathrm{EX}\Gamma_n, \mathrm{AX}\Delta_1, \ldots, \mathrm{AX}\Delta_m, \Lambda}{\mathrm{E}\Gamma_1, \mathrm{A}\Delta_1, \ldots, \mathrm{A}\Delta_m \quad \ldots \quad \mathrm{E}\Gamma_n, \mathrm{A}\Delta_1, \ldots, \mathrm{A}\Delta_m} \quad \mathbf{R}$$

$$\text{(X}^-) \quad \frac{\mathrm{AX}\Delta_1, \ldots, \mathrm{AX}\Delta_m, \Lambda}{\mathrm{A}\Delta_1, \ldots, \mathrm{A}\Delta_m} \qquad \text{(Ett)} \quad \frac{\mathrm{E}\emptyset, \Phi}{\Phi}$$

Fig. 3.16.   Rules for CTL* satisfiability games.

---

[66]Demri *et al.* (2016, p. 695).

play $C_0, C_1, \ldots$ There is a connection relation between B, C_i and B', C_{i+1} if the rule applied between C_i and C_{i+1}

(i) did not operate on B and $B' = B$, or
(ii) transformed B into B'.

A trace in an infinite play $C_0, C_1, \ldots$ of game $\mathcal{G}(\vartheta)$ is an infinite sequence of blocks $B_0, B_1, \ldots$ such that for all $i \in \mathbb{N}$, there is a connection relation between B, C_i and B', C_{i+1}. A trace $Q_0 \Delta_0, Q_1 \Delta_1, \ldots$ is called E-trace if there is some $n \in \mathbb{N}$ with $Q_i = \mathrm{E}$ for all $i \geq n$. In an A-trace, there is some $n \in \mathbb{N}$ with $Q_i = \mathrm{A}$ for all $i \geq n$. With König's lemma, it follows that every infinite play has a trace.

A thread is an infinite trace $B_0, B_1, \ldots$ is an infinite $\varphi_0, \varphi_1, \ldots$ such that there is a connection relation between φ_i, B_i and φ_{i+1}, B_{i+1} for all $i \in \mathbb{N}$. Again, a thread is called a μ-thread if there is some formula ψ of the form $\psi_1 \cup \psi_2$ and some $n \in \mathbb{N}$ with $\varphi_i = \mathrm{X}\psi$ for all $i \geq n$. A thread is called a ν-thread if there is some formula ψ of the form $\psi_1 \mathrm{R} \psi_2$. Every trace has at least one thread. Every thread is either a μ-thread or a ν-thread.

The winning conditions for infinite plays are defined with μ-threads and ν-threads. A trace is called good if

(i) it is an A-trace and contains a ν-thread, or
(ii) it is an E-trace and contains no μ-thread.

V wins an infinite play of the satisfiability game $\mathcal{G}(\vartheta)$ if all its traces are good. **R** wins if it contains a trace which is not good ("bad").

Figure 3.17 illustrates a play in the satisfiability game for the CTL* formula $\vartheta = p \wedge \mathrm{AG}(p \to \mathrm{EXF}p) \wedge \mathrm{AFG}\neg p$. The formula ϑ is

$A(G(\neg p \vee EXFp)),A(FG\neg p),p$
$A(\neg p \vee EXFp),A(XG(\neg p \vee EXFp)),A(G\neg p,XFG\neg p),p$
$A(\neg p,EXFp),A(XG(\neg p \vee EXFp)),A(\neg p,XFG\neg p),A(XG\neg p,XFG\neg p),p$
$E(XFp),A(XG(\neg p \vee EXFp)),A(XFG\neg p),A(XG\neg p,XFG\neg p),p$
$E(Fp),A(G(\neg p)),A(FG\neg p),A(G\neg p,XFG\neg p)$
$E(XFp),A(\neg p,EXFp),A(XG(\neg p \vee EXFp)),A(\neg p,XFG\neg p),A(XG\neg p,XFG\neg p)$
$E(XFp),A(XG(\neg p \vee EXFp)),A(XG\neg p,XFG\neg p),\neg p$
$E(Fp),A(G(\neg p \vee EXFp)),A(G\neg p,FG\neg p)$
$Ep,A(\neg p,EXFp),A(XG(\neg p \vee EXFp)),A(\neg p,XFG\neg p),A(XG\neg p,XFG\neg p)$
$E(XFp),A(XG(\neg p \vee EXFp)),A(XFG\neg p),A(XG\neg p,XFG\neg p),p$
$\vdots$

Fig. 3.17.   Example of a play in a CTL* satisfiability game.[67]

---

[67]Demri *et al.* (2016, p. 698).

unsatisfiable because it states the negation of a valid statement: If p is true now and whenever p is true somewhere, and then it holds again somewhere later, then there must be a path on which it is true infinitely often.

In the play in Fig. 3.17, two rules for the temporal operators G and F are applied, which can easily be derived from the list of rules in Fig. 3.16[68]:

$$
\text{(AF)} \quad \frac{A(F\varphi, \Delta), \Phi}{A(\varphi, XF\varphi, \Delta), \Phi} \qquad \text{(AG)} \quad \frac{A(G\varphi, \Delta), \Phi}{A(\varphi, \Delta), \Phi \; A(XG\varphi, \Delta), \Phi} \quad \mathbf{R}.
$$

The first configuration in the play in Fig. 3.17 is equivalent to ϑ. The derivation from the first configuration to the second one is realized by the application of (AG) and (AF). The next step from the second to the third configuration is done by the rules (A ∨) and (AG). The fourth configuration is derived by rule (AQ). With rule (X), the fifth configuration is derived.

At each stage, after application of rule (X) with a resulting single configuration only, $\mathbf{V}$ chooses whether to have p in the literal part of the configurations. If $\mathbf{V}$ eventually chooses not to have p ever again, then the play has a bad E-trace. If $\mathbf{V}$ chooses to have p in the only infinite play, the resulting play may be such as in Fig. 3.17. The lightly shaded blocks are an A-trace. The darkly shaded formulae in this A-trace generate a μ-thread. Therefore, it is a bad A-trace. In summary, in any case, $\mathbf{R}$ is the winner of this play. Whatever $\mathbf{V}$ will choose, it leads either to contradicting literals or to bad E-traces or A-traces.

The satisfiability games are sound and complete for the full branching-time logic CTL*. Completeness means that for every state formula ϑ of CTL*, if ϑ is satisfiable, the verifier $\mathbf{V}$ has a winning strategy in the game $\mathcal{G}(\vartheta)$. For soundness, it holds that for every state formula ϑ of CTL*, if verifier $\mathbf{V}$ has a winning strategy in the game $G(\vartheta)$, then ϑ is satisfiable.

---

[68]Demri *et al.* (2016, p. 699).

3.5. Dialogue-Based Constructive Temporal Logic

The game-based interpretation of logic is deeply rooted in the tradition of philosophy. In order to find the truth, platonic philosophy started with dialogues between teachers and students in a game of attacks and defenses of arguments. The Socratic method of dialogues had to obey logical rules, which were later systematized in the famous textbooks of Aristotle, the father of logic. But, actually, the origin of logic was and is embedded in the practice of argumentation with lawyers, politicians, and ordinary people, which can be understood as a game-based approach to logic. In the following, we consider a dialogue-based interpretation of logic which can be extended to modal and temporal logic. A dialogue on a statement between a proponent and an opponent is represented in tables which unfold the sequents of formulae in a Gentzen-style manner.

With respect to applications in computer science, we consider intuitionistic, resp. constructive, logic, which was already discussed in Section 1.2. The reason is that linear and branching-time logic is based on time trajectories and trees, which must be constructed and calculated in a computer step by step, such as in intuitionistic mathematics. Therefore, the classical alternative of either true or false cannot be applied to quantified sentences. Classical approaches to dialogue games were favored by Hintikka and Sandu.[69] The following constructive approach was first suggested by Lorenzen and Lorenz.[70]

3.5.1. *Dialogue-based introduction of constructive logical operators*

Each logical operator must be characterized by an attack–defense rule for the proponent and opponent of an assertion. If the proponent asserts a negation $\neg A$, then the opponent responds with a counterattack A with a question mark, ?. Then, the proponent has

---

[69]Hintikka and Sandu (2009, pp. 341–343).
[70]Lorenz (1968, pp. 73–100).

to defend her assertion by proving that the assumption of A leads to a contradiction:

Assertion	Attack	Defense
$\neg\, A$	A ?	

A constructive defense of an existentially quantified assertion $\lor x\, A(x)$ needs a concrete example s (e.g., object, machine state, temporal point):

Assertion	Attack	Defense
$\lor x\, A(x)$	?	$A(s)$

For the attack of an all-quantifier, the opponent chooses any object s:

Assertion	Attack	Defense
$\land x\, A(x)$	s?	$A(s)$

But if an all-sentence has been defended against an attack, the assertion cannot be called true. It must be defended against all possible and allowed attacks in a dialogue. Therefore, at first, the rules of attacking and defending must be defined for all logical operators. Junctors can no longer be defined by truth tables because the connected sentences may also contain quantifiers. The rules for the adjunction $\lor$ are in analogy to the existential quantifier:

Assertion	Attack	Defense
$A \lor B$	?	A
$A \lor B$	?	B

For the conjunction, in analogy to the all-quantifier, the opponent can choose to attack the left part $(L?)$ or the right part $(R?)$:

Assertion	Attack	Defense
$A \land B$	$L?$	A
$A \land B$	$R?$	B

For an attack–defense rule of the junctor $\rightarrow$, the opponent attacks $A \rightarrow B$ by assuming the premise A but requesting a defense of the conclusion B. The defense of the proponent is B:

Assertion	Attack	Defense
$A \rightarrow B$	A ?	B

Obviously, this is a generalization of the negation rule because intuitionistically $\neg A :\leftrightarrow A \rightarrow \perp$. The controverse conditional $\leftarrow$ is defined by $B \leftarrow A :\leftrightarrow B \rightarrow A$. Thus, the corresponding attack–defense rule requests:

Assertion	Attack	Defense
$A \leftarrow B$	B ?	A

The other junctors can be defined by $\wedge, \vee, \rightarrow$, and $\leftarrow$ in the usual way. A dialogue begins with the assertion of thesis which a proponent writes down on the right-hand side of a tableau divided by a double line. Then, the opponent attacks the thesis, and the dialogue continues by means of alternating defenses and attacks until atomic propositions are reached. In the dialogue tableau, verified propositions are enclosed by brackets. An example is the thesis "All politicians are stupid or clever," or in symbols, $\wedge x(p(x) \rightarrow s(x) \vee c(x))$:

		$\wedge x(p(x) \rightarrow s(x) \vee c(x))$
Kennedy ?		$p(K) \rightarrow s(K) \vee c(K)$
$p(K)$	?	?
$[p(K)]$		$s(K) \vee c(K)$
	?	$c(K)$
	?	

If the asserted atomic proposition $c(K)$ is falsified, then the proponent has lost the game. His thesis is not true.

3.5.2. Constructive dialogue games

The first rules of a dialogue game are the following.

Starting rule: The proponent begins by asserting a thesis. The players make their moves alternatively.

General rule: Each player may either attack a sentence asserted by the other player or defend himself against an attack by the other player.

So far, truth means defensibility of a thesis in a dialogue game against any opponent. In the case of sentences without quantifiers, there are always only finitely many strategies for the players. In this special case, it can be decided whether there is a winning strategy for the proponent. Then, the thesis is true. If the atomic propositions are definitely true or false, the notion of truth coincides with the classical notion of truth. The defensibility in the sense of the existence of a winning strategy may not be decidable for quantified sentences.

Furthermore, there may be unclear situations of defense for iterated implications[71]:

$$
\begin{array}{c|ccc}
& ((A \to B) \to C) \to A & \\
(A \to B) \to C \ ? & A \to B & ?2 \\
A & & ?3 \\
{[A]} & &
\end{array}
$$

The proponent could defend her thesis by asserting A. But the opponent could argue that she asserted A only in the attack against $A \to B$. Therefore, the proponent should first defend herself against this attack by asserting B. Consequently, the general rule should be restricted in the following way.

Each player may either attack a sentence asserted by the other player or she may defend herself against the last attack against which she has not already defended herself.

It is a disadvantage of this rule that the opponent could indefinitely repeat her attack and the proponent could never win the game. Therefore, a limit of repetitions seems to be reasonable in the sense that the opponent may attack a sentence at most $m + 1$ times for an arbitrarily chosen m. With this restriction, a winning rule can be defined as follows.

---

[71]Lorenzen (1969, p. 28).

Winning rule: If the proponent cannot make any further move, the proponent has won.

In order to avoid the somewhat artificial choice of a restrictive number of attacks, the general rule can equivalently be changed in the following way:

(i) The proponent may either attack a sentence asserted by the opponent or she may defend himself against the last attack of the opponent.

(ii) The opponent may either attack the sentence asserted by the proponent in the preceding move or she may defend herself against the attack of the proponent in the preceding move.

With this rule, truth is defined as defensibility against every opposition. If the defensibility of a thesis depends on the empirical truths of atomic propositions, then the thesis is called empirically true. For example, consider the thesis "All seas in the world are polluted," or formally, with the atomic proposition $a(y, x)$ ("y is in x"):

$$
\begin{array}{c|c}
\text{Atlantic ?} & \wedge x_{\text{seas}} \vee y_{\text{pollution}}\, a(x, y) \\
? & \vee y_{\text{pollution}}\, a(\text{Atlantic}, y) \\
? & a(\text{Atlantic, oil pollution})
\end{array}
$$

In logic, there are theses which are defensible independent of the meaning of propositions. These are the logically true formulae, for example:

$$
\begin{array}{c|c}
& a \to a \\
a \quad ? &
\end{array}
$$

The opponent would even lose the game if she can defend a as an empirically true proposition. In this case, the proponent could defend himself with the same proposition. Another example is the following logically true formula:[72]

---

[72]Lorenzen (1969, pp. 32–33).

1			$\neg(a \wedge \neg a)$	
2	$a \wedge \neg a$	?	$L\,?\,2$	
3	a		$R\,?\,2$	
4	$\neg a$		a	$?\,4$

In this case, again, the opponent must attack the proposition a which she herself has asserted in line 3. Therefore, she has to give up the game. Another famous example is the modus ponens. In this case, two hypotheses are given, and the thesis has to be defended in such a way that the proponent finally has to defend an elementary proposition which has been asserted by the opponent:

1	$\wedge x(a(x) \to b(x))$			
2	$\wedge x(b(x) \to c(x))$		$\wedge x(a(x) \to c(x))$	
3		$y\,?$	$a(y) \to c(y)$	
4	$a(y)$	?		$y\,?\,1$
5	$a(y) \to b(y)$		$a(y)$	$?\,5$
6	$b(y)$			$y\,?\,2$
7	$b(x) \to c(x)$		$b(y)$	$?\,7$
8	$c(x)$		$c(x)$	

If a thesis can be defended under the assumption of hypotheses, the hypotheses are called to logically imply the thesis. According to formal logic, logical truth and logical implications are studied in a formal variant of dialogical games.

3.5.3.   *Constructive formal dialogue games*[73]

Formal dialogues do not start with true or false propositions but with prime formulae $a, b, \ldots, a(x), b(y), \ldots, a(x.y), \ldots$. The syntax is given by the symbols of operators:

$$\wedge |\vee| \to | \neg | \wedge x| \vee x.$$

The attack–defense rules of the operators are the same as before. For prime formulae p, the following rule with attack (?) and no

---

[73]Lorenzen (1969, pp. 36–37).

defense is supplemented:

$$p|?|.$$

In formal games, the proponent is never allowed to attack a prime formula. Now, the general and winning rules request for formal games are as follows.

General rule for formal games:

(i) The proponent may either attack a composite formula of the opponent or she may defend herself against the last attack of the opponent.

(ii) The opponent may either attack the formula put forward by the proponent in the preceding move, or she may defend herself against the attack of the proponent in the preceding move.

Winning rule for formal games:

If the proponent has to defend a prime formula which has been put forward by the opponent, the proponent has won.

With respect to formal calculi of intuitionistic, resp. constructive, logic, the notion of dialogical defensibility delivers what is wanted. An example is double negation, which is excluded in intuitionistic logic as logically true:

1	$\neg\neg a$		a	
2		$?$	$\neg a$	$?\,1$
3		$a\ ?$		

The proponent has no move besides repeating his attack of $\neg\neg a$. Therefore, he cannot win the game. In classical logic, the tertium non datur $a \vee \neg a$ is assumed. Under this condition, the double negation is formally defensible:

1	$a \vee \neg a$			
2	$\neg\neg a$		a	
3		$?$		$?\,1$
4	$\neg a$		$\neg a$	$?\,2$
5	a	$?$	a	$?\,4$

Obviously, the opponent has to give up.

In classical formal quantified logic, the formula $\neg \wedge x\,a(x)$ logically implies $\vee x\,\neg a(x)$. In a formal dialogue game, $\vee x\,\neg a(x)$ is only defensible with the additional hypotheses $\wedge x(a(x) \vee \neg a(x))$ and $\vee x\,(\neg a(x) \vee \neg \vee x \neg a)$:

1	$\vee x\,(\neg a(x) \vee \neg \vee x \neg a(x))$				
2	$\wedge x(a(x) \vee \neg a(x))$				
3	$\neg \wedge x\,a(x)$		$\vee x \neg a(x)$		
4		?		?1	
5	$\neg \vee x \neg a(x)$		$\wedge x\,a(x)$	?3	
6		y?		y?2	
7	$a(x) \vee \neg a(x)$			?7	
8	$\neg a(y)$		$\vee x\,\neg a(x)$	?5	
9		?	$\neg a(y)$		
10	$a(y)$	?	$a(y)$	?8	

3.5.4. Constructive dialogue-based logic as Gentzen's intuitionistic calculus $G3$[74]

The tableaux of dialogue games can be transformed into a Gentzen-style calculus. Therefore, a dialogue tableau is written in one line as a position:

$$
\begin{array}{c}
A_1 \\
\vdots \\
A_n
\end{array} \Big\| \,B \qquad \text{as} \quad A_1, \ldots, A_n \,\|\, B
$$

A formal calculus can be given which delivers all the winning positions. A formal system of hypotheses is denoted by Σ, and a system Σ in which a formula A occurs by $\Sigma(A)$. For all prime formulae p, there are winning positions $\Sigma(p) \,\|\, p$, which are used as initial positions of the calculus.

In the second step, the rules must be given to derive winning positions from winning positions. One example is the following rule

---

[74]Kleene (1967, §80).

with $\rightarrow$ in the conclusion:

$$\Sigma, A \parallel B \Rightarrow \Sigma \parallel A \rightarrow B$$

Obviously, the opponent has only one attack

$$
\begin{array}{c|c}
\Sigma & A \rightarrow B \\
A? &
\end{array}
$$

Then, there is the following defense:

$$
\begin{array}{c|c}
\Sigma & A \rightarrow B \\
A? & B
\end{array}
$$

The corresponding rule has $\rightarrow$ in the hypotheses:

$$\Sigma(A \rightarrow B) \parallel A; \ \Sigma(A \rightarrow B), B \parallel C \Rightarrow \Sigma(A \rightarrow B) \parallel C$$

In position $\Sigma(A \rightarrow B) \parallel C$, the opponent must attack C. Then, the proponent can attack $A \rightarrow B$:

$$
\begin{array}{c|c}
\Sigma(A \rightarrow B) & C \\
? & A \quad ?
\end{array}
$$

The opponent may only continue with

$$
\begin{array}{c|c}
\Sigma(A \rightarrow B) & C \\
\ldots \quad ? & A \quad ? \\
\ldots \quad ? &
\end{array}
$$

or with

$$
\begin{array}{c|c}
\Sigma(A \rightarrow B) & C \\
\ldots \quad ? & A \quad ? \\
B &
\end{array}
$$

If both positions are winning positions, then $\Sigma(A \rightarrow B) \parallel C$ is also a winning position. In this manner, two rules can be introduced for each logical operator $\neg, \rightarrow, \wedge x$, and $\vee x$. For the junctors $\wedge$ and $\vee$, there are even three rules because either the opponent or the proponent can choose between two attack–defense rules. Thus, there are $4 \cdot 2 + 2 \cdot 3 = 14$ attack–defense rules.

These 14 rules together with the initial positions $\Sigma(p) \parallel p$ form a complete logical calculus: The hypotheses $A_1, \ldots, A_n$ logically imply B if and only if the position $A_1, \ldots, A_n \parallel B$ can be derived in this calculus. This calculus is the intuitionistic Gentzen calculus G3, which is equivalent to Heyting's calculus of intuitionistic logic. Obviously, the Gentzen calculus G3 is an appropriate tool to study formal dialogical games.

3.5.5.   *Dialogue-based modal logic*[75]

The modal operator Δ ("necessity") can be introduced with respect to logical truth and implication. If a system Σ of sentences is accepted as true, a logically implied sentence A (i.e., $\Sigma \models A$) *must* be accepted as true. In this sense, A is said to be necessary relative to Σ. In short: $\Delta_\Sigma A :\leftrightarrow \Sigma \models A$ (with A as a modal-free sentence). If A is logically true independent of the particular system Σ of sentences, it is said to be necessary (i.e., ΔA). As a theorem of logical implication, it can be proven that $\Delta A \wedge \Delta B \to \Delta(A \wedge B)$ is true. With respect to modal logic, it is said that $\Delta A \wedge \Delta B$ modal-logically implies $\Delta(A \wedge B)$ (in short: $\Delta A \wedge \Delta B \models \Delta(A \wedge B)$). Modal logic studies the class of modal-logical implications.

An example is the so-called rule of Aristotle, which can easily be proven with the suggested interpretation of the Δ-operator:

$A_1 \wedge \cdots \wedge A_n$ logically implies $B \Rightarrow \Delta A_1 \wedge \cdots \wedge \Delta A_n$ logically implies ΔB.

The convertible version is also valid:

$\Delta A_1 \wedge \cdots \wedge \Delta A_n$ logically implies $\Delta B \Rightarrow A_1 \wedge \cdots \wedge A_n$ logically implies B.

If $\Delta_\Sigma A_1 \wedge \cdots \wedge \Delta_\Sigma A_n$ logically implies $\Delta_\Sigma B$ for all systems Σ, then, especially, it holds for $\Sigma = A_1 \wedge \cdots \wedge A_n$, which gives B.

Another implication of Aristotle, which can be easily justified in this manner, requests:

ΔA logically implies A.

---

[75]Lorenzen (1969, Chapter 6).

How can a compound sentence be defended as thesis if some hypotheses are given? In a dialogue, the opponent may have put forward a formula ΔA. Without the restriction that Σ is unknown, the proponent could ask, relative to which Σ the necessity is asserted. But, in modal logic, that is not allowed. Thus, the proponent may only force the opponent to admit A if she has admitted ΔA:

$$\Delta A \quad | \quad ?| \quad A$$

If the proponent has put forward a Δ-formula ΔB, and if the opponent attacks with "?," the proponent must defend ΔB as logically implied by all formulae put forward by the opponent beforehand. Only the rule of Aristotle is available for defending ΔB. Therefore, the following rule is supplemented.

Δ-**defense rule:** If the proponent defends a Δ-formula, she may attack only the Δ-formulae (without the beginning Δ) put forward by the opponent beforehand.

The logical Gentzen calculus G3 can be extended to deliver all the winning positions of the dialogical game with Δ. The following rule, which corresponds to "ΔA logically implies A," leads from winning positions to winning positions:

$$(O_\Delta)\, S(\Delta A), A \parallel B \Rightarrow S(\Delta A) \parallel B$$

A system $S(\Delta A)$ of hypotheses with a Δ-formula ΔA is given. In order to defend a thesis B, the proponent may attack ΔA. The opponent must defend herself by putting forward A. Thus, a new position $S(\Delta A), A \parallel B$ arises.

Another rule which leads from winning positions to winning positions is a version of Aristotle's rule:

$$(P_\Delta) \quad A_1, \ldots, A_n \parallel B \Rightarrow S(\Delta A_1, \ldots, \Delta A_n) \parallel \Delta B$$

A system $S(\Delta A_1, \ldots, \Delta A_n)$ of hypotheses is given. In order to defend a thesis ΔB, the proponent may try to defend B with the system $A_1, \ldots, A_n$ of hypotheses.

The Gentzen calculus for modal logic is consistent and complete with respect to the dialogue-based interpretation of Δ.

In the philosophical tradition of Aristotle and Thephrastus, new modalities were introduced for application in syllogisms with terms of the Δ-operator and negation only:

$$\Delta'A: \leftrightarrow \Delta\neg A,$$
$$\nabla'A: \leftrightarrow \neg\Delta A,$$
$$\nabla A: \leftrightarrow \neg\Delta\neg A$$

Δ and Δ' are contraries. Δ, ∇' and Δ', ∇ are contradictories. There-fore, ΔA logically implies ∇A, and $\Delta'A$ logically implies $\nabla'A$, which is illustrated in a modal square as follows.

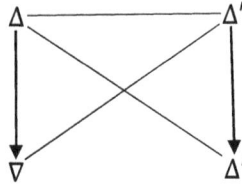

In philosophical tradition, ∇ is understood as "possible," $\bowtie$ $A: \leftrightarrow$ $\nabla A \wedge \nabla'A$ as "contingent." Aristotle and Thephrastus understood truth and falsity also as modalities, which are indicated by the oper-ators X, resp. X'. They can be used to define contingent truth and contingent falsity:

$\bowtie$ $A :\leftrightarrow XA \wedge \nabla'A$ (contingently true),
$\bowtie$ $'A :\leftrightarrow X'A \wedge \nabla A$ (contingently false).

For these new modalities, another modal diagram with implica-tions can be introduced (Fig. 3.18):

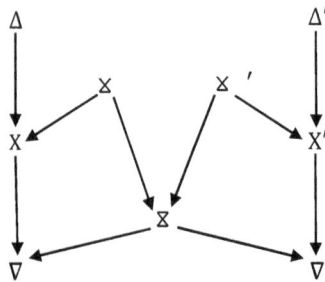

Fig. 3.18. Modal diagram.[76]

---

[76]Lorenzen (1969, p. 67).

3.5.6. *Dialogue-based temporal logic*

Since antiquity, necessary, possible, and contingent truth has been discussed in temporal logic of everyday life and physics. Elementary propositions $a(S, t)$ are related to systems S of physical objects and time t. They are sometimes called empirical or material elementary propositions, which are verified or falsified by observations and measurements. In classical physics, it is assumed that this kind of propositions is either true or false independent of observations and measurements. They are called value-definite. In observational praxis, one has to consider the temporal sequence of observations and measurements. Therefore, in a dialogue of proponent P and opponent O, one has to consider the temporal order of attacks and defenses. An example is the sequential conjunction $a \sqcap b$, which means "a and then b." The attack–defense rule is illustrated in the following dialogue scheme. The concerned line of attack is given in brackets:[77]

As the elementary propositions are assumed to be value-definite, the dialogue can be illustrated in a temporarily ordered tree (Fig 3.19)A. At the nodes, a and b are tested. $\bar{a}$ means that a is false. The arrow marks the temporal sequence of the dialogue.

	O	P
1		$a \sqcap b$
2	1? (0)	a
3	a? (1)	$[a]$
4	2? (0)	b
5	b? (4)	$[b]$

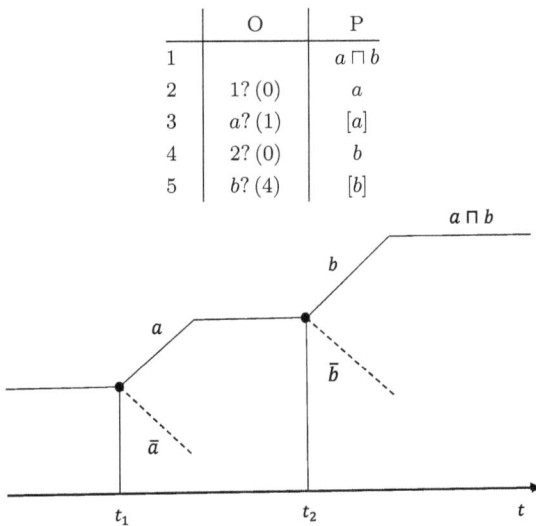

Fig. 3.19.   Temporal tree of consequential conjunction.

[77]Mittelstaedt (1986, p. 47).

The temporal order of tests is uniquely determined. The temporal difference $\delta = t_2 - t_1$ may be arbitrarily small. In the limit $\delta \to 0$, both temporal values coincide. This is the case of logical conjunction. Its dialogue-based scheme was already introduced. Contrary to a dialogue tableau, a temporal tree contains all possible developments which are illustrated for the logical connectives $\wedge, \vee$, and $\neg$ in Fig. 3.20. In this case, it is an idealization of $\delta = t_2 - t_1 \to 0$.

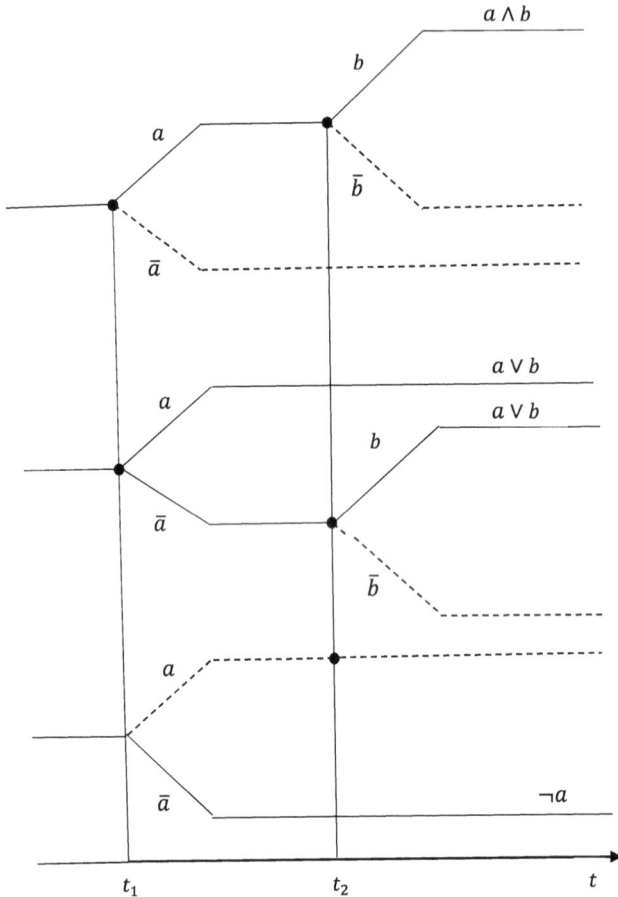

Fig. 3.20.   Temporal trees of logical connectives.[78]

---

[78]Mittelstaedt (1986, p. 53).

In philosophical tradition, the assumptions of classical physics were more or less understood as self-evident conditions of temporal logic. But, actually, the concepts must be changed with respect to modern physics, which will be considered in Section 4.2.

Chapter 4

Applications and Outlook
of Temporal Logic

4.1. Temporal Logic in Artificial Intelligence and Machine Learning

Formal reasoning in temporal logic delivers efficient tools to control and certify data-driven learning procedures in artificial intelligence (AI). Because of an explosion of data in practical applications of machine learning, AI is endangered by biases and failures without these certifications. This chapter starts with a survey on the development of AI: The emergence of three paradigms can be distinguished — symbolic, subsymbolic, and hybrid AI (Fig. 4.1). In the following, the calculi of temporal logic will be combined as tools of symbolic AI with subsymbolic AI in the new promising paradigm of hybrid AI.

4.1.1. *Symbolic AI*

The first paradigm of AI has been dominated by symbolic logic, and therefore, it is called symbolic AI. In the first phase of AI research, the search for general problem-solving methods was successful at least in formal logic.[1] J. A. Robinson proposed the so-called resolution method, according to which proofs can be found by logical

---

[1]Mainzer (1981, pp. 22–33, 133–134).

degrees of intelligence

hybrid AI

hybrid cognitive
systems

combination of machine
learning with automatic proving
and logical methods

subsymbolic AI

sensor systems
("perception")

machine learning with
big data

symbolic AI

logical systems
("reason")

automatic proving and
knowledge-based systems

Turing test 1950

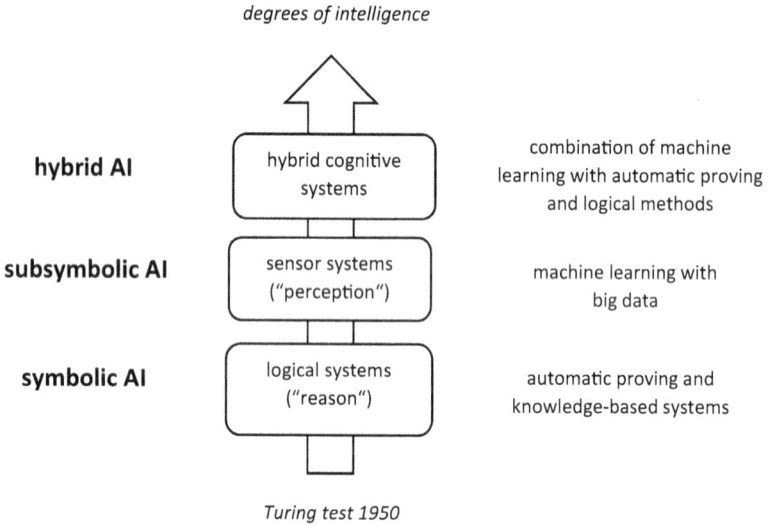

Fig. 4.1. From symbolic and subsymbolic to hybrid AI.

refutation procedures.[2] Automatic reasoning with logical procedures of satisfiability (SAT) opened practical applications of SAT-solving, which are still used in industrial logistics (e.g., car production).[3]

To solve a problem with a computer, the problem must be translated into a programming language. A corresponding programming language is called "Programming in Logic" (PROLOG), which has been in use since the early 1970s.[4] As an alternative to statements and relations, knowledge can also be represented by functions and classifications, such as those used in mathematics. Functional programming languages, therefore, do not regard programs as systems of facts and conclusions (such as PROLOG) but as functions of sets of input values in sets of output values.[5] One example is the functional programming language LISP, which was developed by J. McCarthy as early as the end of the 1950s during the first AI phase (McCarthy, 1960).

---

[2]Robinson (1965, pp. 23–41), Mainzer (1985, pp. 41–56).
[3]Biere *et al.* (2009).
[4]Hanus (1986).
[5]Church (1941).

In LISP, knowledge is represented by data structures and knowledge processing by algorithms as effective functions. Knowledge-based expert systems are computer programs that store and accumulate knowledge about a specific area, from which knowledge automatically draw conclusions in order to offer solutions to concrete problems in that area. In contrast to a human expert, the knowledge of an expert system is limited to a specialized information base without general and structural knowledge about the world.[6]

In 1969, the logician Howard observed that Gentzen's proof system of natural deduction can be directly interpreted in its intuitionistic version as a typed variant of the mode of computation known as λ–calculus.[7] This basic insight of mathematical proof theory opened avenues to new generations of interactive and automated proof assistants. The proof assistant Coq is an example, which is based on the calculus of inductive constructions (CiC), combining both a higher-order logic and a richly typed functional language.[8] There are more proof assistants, such as Agda and Isabelle. From a practical point of view, the disadvantage is the increasing complexity of AI programs, which often cannot be handled by proof assistants with the accuracy of mathematical proofs. Against the background of earlier studies, practical limits of verification and certification in symbolic AI by proof assistants were studied recently.[9]

4.1.2. Subsymbolic AI

Skills and intuition of human experts cannot be completely grasped by formal knowledge-based systems in symbolic AI. The second paradigm of AI is dominated by learning from Big Data of experience beyond the symbolic logic paradigm, and therefore, it is called subsymbolic AI.

Statistical learning tries to derive a probabilistic model from a finite amount of data in experience (e.g., observations and experiments) by algorithms. Statistical reasoning attempts, conversely, to derive properties of observed data from an assumed statistical

---

[6]Puppe (1988), Mainzer (1990).
[7]Howard (1969).
[8]Bertot and Castéran (2004).
[9]Mainzer *et al.* (2018), Mainzer and Kahle (2021).

model by algorithms. Learning algorithms of machine learning[10] are based on statistical learning theory.[11] The paradigm of formal logic is replaced by the paradigm of statistics and probability theory. The limits of probabilistic reasoning and learning stand in a long tradition of epistemic debates since Hume and Kant.

Learning algorithms can be realized by technical neural networks.[12] They are complex dynamical systems of firing and non-firing neurons with (simplified) topologies like those in living brains. The dynamics of neural nets can be modeled in phase spaces of synaptic weights with trajectories converging to attractors which represent prototypes of patterns.[13]

4.1.2.1. *Example: Neural networks*

In feed-forward neural networks, neurons are represented as nodes of a graph in layers (Fig. 4.2). In this case, each neuron of a layer is connected with all neurons of the following layer through directed edges (synapses). The neurons of a layer are not connected mutually. Exceptions are the input layer with neurons without input connections and the output layer without output connections. The layers between the input and output layers are called "hidden." In the graphic model of a neural network, each connection or edge is weighted with a number which corresponds to the intensity of the synaptic connection. Each neuron is characterized by an activation function which defines the input–output relation for that neuron.

Mathematically, neural networks can be defined as functions $\nu : I^n \to O^m$ which map a n-dimensional input domain $I^n (n > 0)$ on a m-dimensional output domain $O^m (m > 0)$.[14] This kind of network can compute, for example, an approximation of a function $f : \mathbb{R}^n \to \mathbb{R}$ with $I = O = \mathbb{R}$. A network which classifies, for example, 8-bit pictures of size $h \times v$ (with h horizontal and v vertical) in two classes, can be defined by a function $\nu : I^{h \cdot v} \to O$ with input domain $I = \{0, \ldots, 255\}$ for $2^8 = 256$ possible 8-bit pictures

---

[10]Schölkopf and Smola (2002).
[11]Vapnik (1998).
[12]Schmidhuber (2015, pp. 61, 85–117).
[13]Mainzer (2002).
[14]Mainzer and Kahle (2022); Leofante (2018, p. 2).

output layer y_1 y_2

layer 3

layer 2

layer 1

input layer x_1 x_2

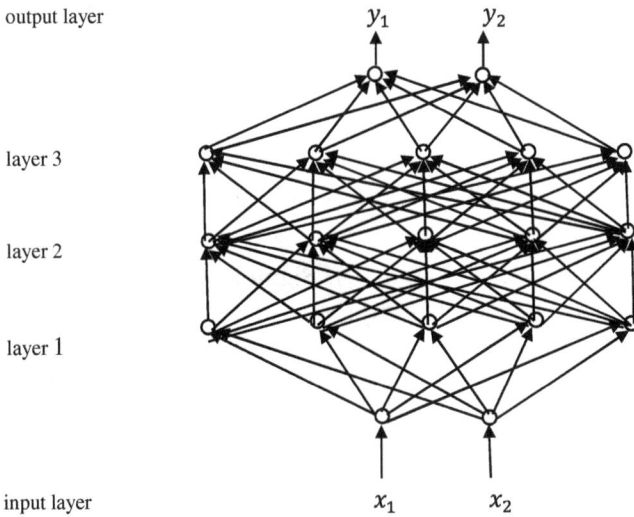

Fig. 4.2. Feedforward network with three hidden layers and an input and output layer.

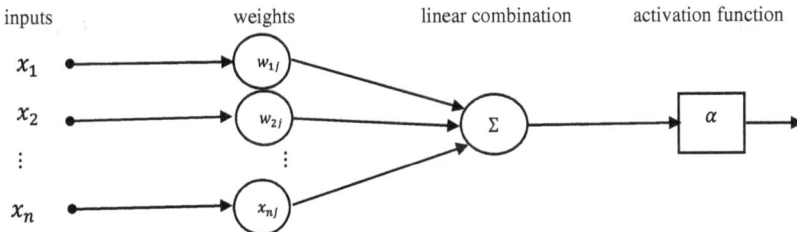

inputs   weights   linear combination   activation function

x_1 w_{1j}

x_2 w_{2j} Σ α

$\vdots$ $\vdots$

x_n x_{nj}

Fig. 4.3. Activation of neurons.

and output domain $O = \{0, 1\}$ for two classes denoted 0 and 1. The mapping starts with an input from I^n, which is at first given to the input layer and then through the hidden layers to the output layer. Mathematically, linear combinations of the values of nodes (neurons) and weights (synaptic connections) from the preceding layers are computed layer by layer. The activation functions are applied on these results for the following neurons (Fig. 4.3).

Networks are distinguished by different activation functions. The threshold function of a McCulloch–Pitts neuron has only the function value of 1 for inputs $v \geq 0$, otherwise 0 (Fig. 4.4(a)). A piecewise linear function maps a bounded interval linearly and the outer

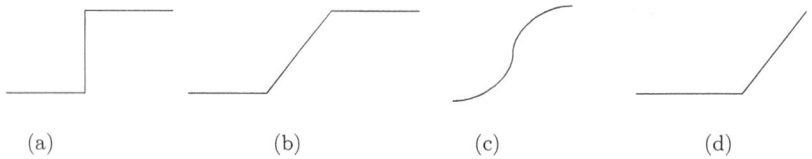

(a)   (b)   (c)   (d)

Fig. 4.4.   Examples of activation functions.

intervals constantly (Fig. 4.4(b)). Sigmoid functions have a variable gradient measure, which is expressed by the curvature of the graph (Fig. 4.4(c)). A rectifier function (ReLU: rectifier linear unit) has the positive values of their arguments, otherwise 0 (Fig. 4.4(d)).

4.1.3. *Statistical and causal learning*

Because of Big Data, neural networks with deep learning are endangered to become a black box with an ever-increasing number of parameters.[15] Data correlations can provide indications of facts but do not have to do so. Statistical learning and reasoning from data is therefore not sufficient. Rather, one must recognize the causal relationships between causes and effects behind the measured data. These causal relationships depend on the laws of the respective application domain. Therefore, in addition to the statistics of the data, additional laws and structural assumptions of the application domains are required, which are verified by experiments and interventions. Causal explanatory models (e.g., the planetary model or a tumor model) fulfill the law and structural assumptions of a theory (e.g., Newton's gravitational theory or the laws of cell biology). In causal reasoning, the properties of data and observations are derived from causal models. Causal inference thus makes it possible to determine the effects of interventions or data changes (e.g., through experiments). Causal learning, vice versa, tries to create a causal model from observations, measurement data, and interventions (e.g., experiments), which require additional laws and structural assumptions.[16]

It can be proved that a causal model includes a clear probability distribution of the data but not vice versa: For causal models

---

[15]Knight (2017).
[16]Pearl (2009), Peters *et al.* (2017).

(e.g., the planetary model), additional laws (e.g., the gravitational law) must be assumed.[17] The objective of causal learning is therefore to discover the causal dependencies of causes and effects behind the distribution of measurement and observation data.

4.1.3.1. *Example: Reinforcement learning*

In unsupervised learning, the algorithm learns to recognize new patterns (correlations) from the set of inputs without "teachers." In supervised learning, the algorithm learns to determine a function from given pairs of inputs and outputs (training). A "teacher" (e.g., trained prototype of a pattern) corrects deviations from the correct function value to an output (e.g., recognition of learned patterns). Reinforcement learning (RL) is in between: A robot is given a target (as in supervised learning). However, it must find the realization independently (as in unsupervised learning). In the step-by-step realization of the goal, the robot receives feedback from the environment at each partial step as to how good or bad it is at realizing the goal. Its strategy is to optimize this feedback.

Technically, this means the algorithm learns through experience (trial and error) how to act in an (unknown) environment (world) in order to maximize the utility of the agent.[18] Mathematically, RL is a dynamic system of an agent and its environment with discrete time steps $t = 0, 1, 2, \ldots$. At any time t, the world is in a state z_t. The agent chooses an action a_t. Then, the system enters the state z_{t+1} and the agent receives the reward b_t (Fig. 4.5).[19]

The agent's strategy is defined by π_t, wherein $\pi_t(z, a)$ is the probability that the action is $a_t = a$ if the state is $z_t = z$. Algorithms of RL determine how an agent changes its strategy based on its experiences (rewards). The goal of the agent is to optimize his feedback in order to achieve the goal.

An example is a mobile robot which is supposed to pick up empty beverage cans in an office and throw them in a trash can. The robot has sensors to detect the cans and an arm with a gripper to grip the cans. Its activities depend on a battery that occasionally needs to be

---

[17]Mooij *et al.* (2013).
[18]Sutton and Barto (1998), Russell and Norvig (2004).
[19]Mainzer (2019, p. 121).

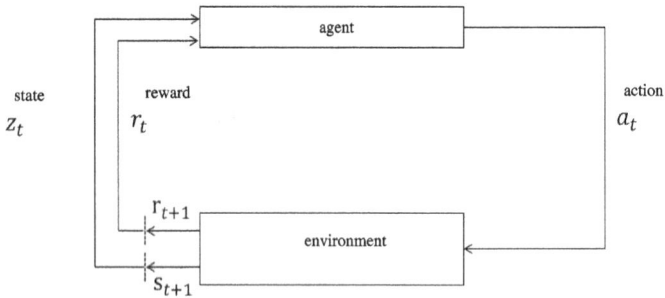

Fig. 4.5.　Reinforcement learning of an agent from its environment.

recharged at a base station. The control system of the robot consists of components for the interpretation of sensor information and for the navigation of the robot arm and robot gripper. The intelligent decisions for can search are realized by a reinforcement algorithm that takes into account the charge level of the battery.

The robot can choose between three actions: active search for a can in a certain time period, stationary hibernation and waiting for someone to bring a can, and returning to base station to recharge the battery. A decision is made either periodically or whenever certain events, such as finding a can, occur. The condition of the robot is determined by the condition of its battery. The rewards are usually zero but become positive when the robot finds an empty can or negative when the battery charge runs out.

Ideally, an agent is in a state that sums up all the past experiences necessary to achieve its goal. Normally its immediate and present perceptions are not sufficient for this. But the complete history of all past perceptions is also not necessary. For the future flight of a ball, it is sufficient to know its current position and speed. It is not necessary to know its complete previous course. In such cases, the history of the present state has no influence on future development. If the probability of a state depends only on the preceding state and a preceding action of the agent in that state, the decision process satisfies the Markov property.

The Markov decision process (MDP) is determined by the Markov property:

$$P(z_{t+1}, r_{t+1} | z_{0:t}, a_{0:t}, b_{0:t}) = P(z_{t+1}, r_{t+1} | z_t, a_t).$$

The action model $P(z_{t+1}|z_t, a_t)$ is the conditional probability distribution that the world changes from state z_t to state z_{t+1} if the agent selects the action a_t; r_{t+1} is the expected return in the next step. Because computing and storage capacities are scarce and costly, practical RL applications often require the Markov feature. Even if the knowledge of the present state is not sufficient, an approximation of the Markov property is favorable. For very large ("infinite") state spaces, the utility function of an agent must be approximated (e.g., state–action–reward–state algorithms (SARSA), temporal difference learning, Monte Carlo methods, dynamic programming).

4.1.4. *Hybrid AI*

Increasing computational power and acceleration of communication need improved consumption of energy, better batteries, miniaturization of appliances, and refinement of display and sensor technology. Under these conditions, intelligent functions can be distributed in a complex network with many multimedia terminals.[20] Together with satellite technology and global positioning system (GPS), electronically connected societies are transformed into cyberphysical systems.[21] They are a kind of symbiosis of man, society, and machine with digital and analog interfaces, which leads to "hybrid AI."[22] Knowledge-based systems (symbolic AI) are combined with machine learning (subsymbolic AI). Communication is not only realized between human partners with ("analog") natural languages but in digital codes with the things of this world. Cyberphysical systems also mean a transformation into an Internet of Things (IoT). Things on the Internet become locally active agents of complex dynamical systems, i.e., communication networks.

Examples of locally active agents are robots combining AI functions with knowledge-based programming and situational learning. Only in this way, it will be possible for robots not only to skillfully coordinate their actions with each other but also to plan and decide how more sophisticated biological systems can be. Neuromorphic

---

[20]Mainzer (2017).
[21]Mainzer (2015).
[22]Mainzer (2016).

computer structures are designed, which do not occur in this way in nature but combine the advantages of neuronal systems of nature with the advantages of technical computer structures.[23] Neural quantum computers are also conceivable, in which the enormous computing speed and storage capacity of quantum computers are combined with neural networks.[24] According to this, intelligence is an interactively developing ability and not a static and rigidly programmed property of an isolated system.

But how can one rely on cyberphysical systems with hybrid AI? Statistical machine learning with neural networks works, but we cannot understand and control the processes in the neural networks in detail. Today's machine learning techniques are mostly based only on statistical learning, but that is not enough for safety-critical systems.[25] Machine learning can be combined with proof assistants (symbolic AI) and causal learning. As in cognitive systems, experience and learning from data need control over logical reasoning.

Verification of software has already been a crucial step in the development of computer programs in software engineering[26]: After requirement engineering, design, and implementation of a program, different verification procedures have been applied in practice. A program is said to be correct ("certified") if it can be verified to follow a given specification which is derived from the design of the program. In symbolic AI, proof assistants were already considered, which prove the correctness of a computer program in a consistent formalism like a constructive proof in mathematics (e.g., Coq, Agda, MinLog).[27] Obviously, proof assistants are the best formal verification of correctness for certified programs.

But, in industry and business practices, proof assistants seem to be too ambitious because of the increasing complexity of AI software. Therefore, industrial production is often content with *ad hoc* tests or empirical test procedures, which try to find statistical correlations and patterns of failures.[28] A challenge is the verification of machine

---

[23]Mainzer and Chua (2013).
[24]Mainzer (2020).
[25]Klüppelberg *et al.* (2014).
[26]Bourque and Dupuis (2004).
[27]Mainzer *et al.* (2021).
[28]Bertolino (2000), Tretmans and Brinksma (2003).

learning with neural networks and learning algorithms. The increasing complexity of neural networks with an ever-increasing number of parameters generate black boxes which seem only to be trainable by Big Data and testable by *ad hoc* and empirical procedures.

But, in practice, we must also consider the costs of testing. Formal proofs of complex software need an immense amount of time and man power. On the other hand, it is risky to rely on *ad hoc* testing and empirical testing only in safety-critical systems. For certification of AI programs, we must aim at increasing accuracy, security, and trust in software in spite of increasing complexity of civil and industrial applications but with respect to the costs of testing (e.g., utility functions for trade-off time of delivery vs. market value, cost/effectiveness ratio of availability). There is no free lunch for the demands of safety and security. Against the background of earlier studies on interdisciplinary risk and AI standardization,[29] one should aim at a scale of sustainable degrees of certification for responsible and sustainable AI.

4.1.5. *Reinforcement learning with temporal logic*

A widely used method of machine learning is the RL method. The idea is that an agent or robot searches a problem solution in a certain environment and gets rewards to improve its search in learning steps. Thus, RL is an optimization process of problem-solving, which mainly uses a stochastic MDP to converge to the problem solution.

Real-world applications are very complex and cannot be realized in simple trial-and-error procedures. An example is the task of a humanoid robot to learn how to drive a car.[30] A reward function may be the travel duration to a destination. But there are a huge number of details which must also be considered, such as applications of gas and brake, using the steering wheel and transmission, safety requests, and, last but not least, reaching the destination on time. Humans do not learn in a stupid, brute-force manner by testing all possibilities of behavior but reduce complexity through incorporated rules which encourage correct behaviors and penalize risky ones. They aim at policies to maximize the reward functions.

---

[29]Wahlster and Winterhalter (2020).
[30]Li *et al.* (2017).

The incorporated rules of a reward function are called the specification of a task. The temporal coordination of these behaviors is expressed by modal and temporal operators: One must do this request before that request in order to fulfill a certain intention and to avoid a certain obstacle which could eventually emerge under certain constraints, etc. Therefore, it is convenient to use formulae of a temporal logic to describe the specification of a task. A temporal calculus can be used as a formal specification language to express the requirements of what an agent or robot should do. A quantitative semantics of the temporal calculus is used to transform temporal logical formula into (real-valued) reward functions. Increasing rewards transform to better satisfaction of the specifications. It can be proven that better policies of behavior are learned faster by using rewards of a temporal logic than by heuristic procedures of trial and error.

4.1.6. *Policy search in reinforcement learning*

The search for satisfactory policies in RL is made precise in the framework of MDP.[31] An (infinite) MDP is a tuple (S, A, p, R) with continuous set $S \subseteq \mathbb{R}^n$ of states, continuous set $A \subseteq \mathbb{R}^m$ of actions, transition probability function $p : S \times A \times S \rightarrow [0, 1]$ such that $p(s, a, s')$ is the probability of taking action $a \in A$ at state $s \in S$ ending up in state $s' \in S$ or conditional probability $p(s'|s, a)$, and reward function $R : r \rightarrow \mathbb{R}$ on state–action trajectory $r = (s_0, \ldots, s_T)$.

In RL, the transition function p is not known by the learning robot or agent. The reward function can be defined or learned. The goal of RL is to find an optimal stochastic policy $\pi^* : S \times A \rightarrow [0, 1]$ which maximizes the expected accumulated reward:

$$\pi^* = \arg\max_\pi \mathbb{E}_{p^\pi(r)}(R(r)),$$

with the trajectory distribution $p^\pi(r)$, which results from following policy π and reward $R(r)$ for given r.

A policy can be represented by a parameterized model of, for example, a neural network π_θ with the set θ of weights as model

---

[31]Deisenroth (2013); Li *et al.* (2017).

parameters. The policy search tries to find the optimal set of θ with

$$\theta^* = \arg\max_\theta \mathbb{E}_{p^{\pi_\theta}(r)}(R(r)).$$

4.1.7. *Truncated linear temporal logic for reinforcement learning*[32]

A convenient formal language in RL should have the ability to transform the specification of a learning task into a reward function. In the work of Li *et al.* (2017), the formulae of a linear temporal logic (LTL) refer to predicates of functional constraints $f(s) < c$ with real function $f : \mathbb{R}^n \to \mathbb{R}$ and constant c. The syntax of formulae in this temporal logic is defined inductively as usual:

$$\varphi := \top \,|\, f(s) < c \,|\, \neg\varphi \,|\, \varphi \wedge \psi \,|\, \varphi \vee \psi \,|\, \Diamond\varphi \,|\, \Box\varphi \,|\, \varphi \bigcup \psi \,|\, \varphi T \psi \,|\, X\varphi \,|\, \varphi \Rightarrow \psi,$$

with predicate $f(s) < c$ and Boolean connectives and temporal operators $\Diamond$ (eventually), $\Box$ (always), $\bigcup$ (until), T (then), X (next), and $\Rightarrow$ (implication).

These formulae are evaluated with respect to finite time sequences over a set S of states which are generated by an MDP. A sequence of states from time t to time $t + k$ is denoted by $s_{t:t+k}$. The semantics of the formulae is defined in the following way:

$$s_{t:t+k} \models f(s) < c :\Leftrightarrow f(s_t) < c,$$
$$s_{t:t+k} \models \neg\varphi \qquad :\Leftrightarrow \neg(s_{t:t+k} \models \varphi),$$
$$s_{t:t+k} \models \varphi \Rightarrow \psi \quad :\Leftrightarrow (s_{t:t+k} \models \varphi) \Rightarrow (s_{t:t+k} \models \psi),$$
$$s_{t:t+k} \models \varphi \wedge \psi \quad :\Leftrightarrow (s_{t:t+k} \models \varphi) \wedge (s_{t:t+k} \models \psi),$$
$$s_{t:t+k} \models \varphi \vee \psi \quad :\Leftrightarrow (s_{t:t+k} \models \varphi) \vee (s_{t:t+k} \models \psi),$$
$$s_{t:t+k} \models X\varphi \qquad :\Leftrightarrow (s_{t:t+k} \models \varphi) \wedge (k > 0),$$
$$s_{t:t+k} \models \Box\varphi \qquad :\Leftrightarrow \bigwedge t' \in [t, t + k) s_{t':t+k} \models \varphi,$$

---

[32]Li *et al.* (2017).

$$s_{t:t+k} \models \Diamond\varphi \quad :\Leftrightarrow \bigvee t' \in [t, t+k) s_{t':t+k} \models \varphi,$$

$$s_{t:t+k} \models \varphi \bigsqcup \psi :\Leftrightarrow \bigvee t' \in [t, t+k) s_{t':t+k} \models \psi$$
$$\wedge \bigwedge t'' \in [t, t') s_{t'':t+k} \models \varphi,$$

$$s_{t:t+k} \models \varphi T \psi \quad :\Leftrightarrow \bigvee t' \in [t, t+k) s_{t':t+k} \models \psi$$
$$\wedge \bigvee t'' \in [t, t') s_{t'':t+k} \models \varphi.$$

According to these definitions, the temporal operators have the following meaning:

$s_{t:t+k} \models \Box\varphi$, i.e., the specification defined by φ is always satisfied in the state trajectory $s_{t:t+k}$ if the specification of φ is satisfied for every subtrajectory $s_{t':t+k}$ with $t' \in [t, t+k)$.

$s_{t:t+k} \models \Diamond\varphi$, i.e., the specification defined by φ is eventually satisfied in the state trajectory $s_{t:t+k}$ if the specification of φ is satisfied for at least one subtrajectory $s_{t':t+k}$ with $t' \in [t, t+k)$.

$s_{t:t+k} \models \varphi \bigsqcup \psi$, i.e., the specification defined by φ "until" ψ is satisfied in the state trajectory $s_{t:t+k}$ if the specification of φ is satisfied at every time step before ψ is satisfied, and ψ is satisfied at a time between t and $t + k$.

$s_{t:t+k} \models \varphi T \psi$, i.e., the specification defined by φ "then" ψ is satisfied in the state trajectory $s_{t:t+k}$ if the specification of φ is satisfied at least once before ψ is satisfied between t and $t + k$.

A trajectory s of duration k is said to satisfy a formula φ if $s_{0:k} \models \varphi$.

For RL, a specification of a formula φ must be transformed into a real-valued function that can be used as reward. Therefore, a real-valued function $\rho(s_{t:t+k}, \varphi)$ depending on state trajectory $s_{t:t+k}$ and specification formula φ should measure how far $s_{t:t+k}$ is from satisfying or violating the specification φ. The results of ρ are sometimes called robustness degrees. They define the quantitative semantics of the considered LTL calculus:

$$\rho(s_{t:t+k}, \top) := \rho_{\max} \text{ (maximal robustness degree)},$$

$$\rho(s_{t:t+k}, f(s_t) < c) := c - f(s_t),$$

$$\rho(s_{t:t+k}, \neg\varphi) := -\rho(s_{t:t+k}, \varphi),$$

$$\rho(s_{t:t+k}, \varphi \Rightarrow \psi) := \max(-\rho(s_{t:t+k}, \varphi), \rho(s_{t:t+k}, \psi)),$$

$$\rho(s_{t:t+k}, \varphi \wedge \psi) := \min(\rho(s_{t:t+k}, \varphi), \rho(s_{t:t+k}, \psi)),$$

$$\rho(s_{t:t+k}, \varphi \vee \psi) := \max(\rho(s_{t:t+k}, \varphi), \rho(s_{t:t+k}, \psi)),$$

$$\rho(s_{t:t+k}, X\varphi) := \rho(s_{t+1:t+k}, \varphi) \quad (k > 0),$$

$$\rho(s_{t:t+k}, \Box\varphi) := \min_{t' \in [t,t+k)} \rho(s_{t':t+k}, \varphi),$$

$$\rho(s_{t:t+k}, \Diamond\varphi) := \max_{t' \in [t,t+k)} \rho(s_{t':t+k}, \varphi),$$

$$\rho(s_{t:t+k}, \varphi \bigcup \psi) := \max_{t' \in [t,t+k)} \left(\min(\rho(s_{t':t+k}, \psi), \min_{t'' \in [t,t')} \rho(s_{t'':t'}, \varphi)) \right),$$

$$\rho(s_{t:t+k}, \varphi T\psi) := \max_{t' \in [t,t+k)} \left(\min(\rho(s_{t':t+k}, \psi), \max_{t'' \in [t,t')} \rho(s_{t'':t'}, \varphi)) \right).$$

Obviously, $\rho(s_{t:t+k}, \varphi) > 0$ implies $s_{t:t+k} \models \varphi$, and $\rho(s_{t:t+k}, \varphi) < 0$ implies $s_{t:t+k} \not\models \varphi$. An example illustrates the computation of the robustness degrees of specifications. The formula $\varphi = \Diamond(s < 10)$ is a specification with a one-dimensional state s and a state trajectory $s_{0:2} = s_0 s_1 = [11.5]$. The robustness degree is $\rho(s_{0:1}, \varphi) = \max_{t \in \{0,1\}}(10 - s_t) = \max(-1, 5) = 5$. In this case, $\rho(s_t, \varphi) > 0$ implies $s_{0:1} \models \varphi$ and $\rho(s_t, \varphi) = 5$ is a measure for the margin of satisfaction.

With respect to tasks of reinforcement learning, it is convenient that a calculus of temporal logic L is defined over predicates which specify tasks as functions of states. A quantitative semantics of the calculus should introduce a continuous (real-valued) measure of satisfaction. It is also convenient that the specification formulae are evaluated over finite-state trajectories of variable length. This procedure allows per-step evaluation according to the empirically available data. Temporal operators can have time bounds but must not have them. These criteria are fulfilled by the previous calculus. Therefore, this calculus is called truncated linear temporal logic (TLTL). There are other LTLs with different advantages and disadvantages, such as signal temporal logic (STL),[33] metric temporal logic (MTL), bounded temporal logic,[34] and linear temporal logic on finite traces (LTL_f).[35]

4.1.7.1. *Example: Goal reaching while avoiding obstacles*[36]

In a first task, an end-effector should reach the goal position g but simultaneously avoid two obstacles o_1 and o_1. The discrete and continuous rewards are given by

$$
r_1^{\text{discrete}} = \begin{cases} 5, & \text{if } d_g \leq 0.2, \\ -2, & \text{if } d_{o1,2} \leq r_{o1,2}, \\ 0, & \text{otherwise}, \end{cases}
$$

$$
r_1^{\text{continuous}} = -c_1 d_g + c_2 \sum_{i=1}^{2} d_{o_i},
$$

with Euclidean distance d_g between the end-effector and the goal, distance d_{o_i} between the end-effector and obstacle i, and radius r_{o_i} of obstacle i.

The specification of the task is described by a TLTL formula φ_1, which requires eventually to always stay at goal g and always stay away from obstacles, i.e.,

$$
\varphi_1 = \Diamond \Box (d_g < 0.2) \wedge \Box (d_{o_1} > r_{o_1} \wedge d_{o_2} > r_{o_2}).
$$

The resulting robustness degrees are computed by

$$
\rho_1(\varphi_1, (x_e, y_e)_{0:T}) = \min \left(\max_{t \in [0,T)} \left(\min_{t' \in [t,T)} \left(0.2 - d_g^t \right) \right), \right.
$$

$$
\left. \min_{t \in [t,T)} \left(d_{o_1}^t - r_{o_1}, d_{o_2}^t - r_{o_2} \right) \right),
$$

with trajectory $(x_e, y_e)_{0:T}$ of the end-effector position and distance d^t at time t. The robustness degrees are automatically generated from the quantitative semantics. Thus, it is only necessary to specify the TLTL formula φ_1.

---

[33] Donzé and Maler (2010).
[34] Latvala *et al.* (2004).
[35] De Giacomo and Vardi (2013).
[36] Li *et al.* (2017, pp. 4–5).

4.1.7.2. Example: Pick-and-placement tasks of robots[37]

In the next task, a robot should place a piece of bread in a toaster. The gripper position of the robot ranges continuously from 0 to 100, with 0 being the fully closed position. The 21-dimensional state feature space contains seven joint angles and joint velocities, the pose of the end-effector, and the gripper position. The eight-dimensional action space includes seven joint velocities and the desired gripper position. Actions are sent at 20 Hz.

The specification φ_2 of the task requires that the robot always not hit the table or the toaster, eventually reaches the slot, and keeps the gripper closed until the slot is reached, and if the slot is always reached, it implies next that the gripper remains open always, i.e., the TLTL formula:

$$\varphi_2 = \Box\left(\neg(\psi_{\text{table}} \vee \psi_{\text{toaster}})\right) \wedge \Diamond(\psi_{\text{slot}}) \wedge (\psi_{gc} \bigcup \psi_{\text{slot}})$$
$$\wedge \Box(\psi_{\text{slot}} \Rightarrow X \Box (\psi_{go})),$$

with predicates ψ_{table}, ψ_{toaster}, and ψ_{slot} which describe the regions with spatial constraints $(x_{\min} < x_e < x_{\max}) \wedge (y_{\min} < y_e < y_{\max}) \wedge (z_{\min} < z_e < z_{\max})$ and position (x_e, y_e, z_e) of the end-effector.[38] Orientation constraints are specified in a similar way. The predicates ψ_{gc} and ψ_{go} describe the conditions for the gripper to be closed and opened, respectively.

The resulting robustness degree for the specification φ_2 is given by

$$\rho_2(\varphi_2, p_{0:T}^e) = \min\left(\min_{t\in[0,T)}(\max(-\rho_2(\psi_{\text{table}}, p_{t:T}^e), -\rho_2(\psi_{\text{toaster}}, p_{t:T}^e),\right.$$
$$\max_{t\in[0,T)}\rho_2(\psi_{\text{slot}}, p_{t:T}^e), \max_{t\in[0,T)}\left(\min(\rho_2(\psi_{\text{slot}}, p_{t:T}^e),\right.$$
$$\left.\min_{t'\in[0,T)}\rho_2(\psi_{gc}, p_{t':t}^e)\right), \min_{t\in[0,T)}\left(\max(-\rho_2(\psi_{\text{slot}}, p_{t:T}^e),\right.$$
$$\left.\left.\left.\min_{t'\in[t+1,T)}\rho_2(\psi_{go}, p_{t':T}^e))\right)\right).$$

---

[37]Levine *et al.* (2016), Gu *et al.* (2016).
[38]Li *et al.* (2017, pp. 6–7).

The robustness degrees are automatically generated by the quantitative semantics if specification φ_2 is given. If $\rho_2(\varphi_2, p_{0:T}^e) > 0$, then specification φ_2 is satisfied.

4.1.8. *Temporal logic LTL for reinforcement learning*

The syntax of LTL is defined inductively as usual[39]:

$$\varphi := p \,|\, \neg\varphi \,|\, \varphi_1 \wedge \varphi_2 \,|\, X\varphi \,|\, \varphi_1 \bigcup \varphi_2,$$

with atomic proposition p. The other Boolean connectives are defined as usual in classical logic with $\neg$ and $\wedge$. The logical constants true and false can be defined as $\top := p \vee \neg p$ and $\bot := \neg\top$. The temporal modalities "eventually" and "always" can be defined with the "until" operator as $\Diamond\varphi := \top \bigcup \varphi$ and $\Box\varphi := \neg \Diamond \neg\varphi$, respectively. The LTL formula $\Box\Diamond\varphi$ states that $\Diamond\varphi$ will always be true, or "eventually φ," is repeated forever. In this case, φ must be true infinitely many times. Therefore, $\Box\Diamond\varphi$ means "infinitely often φ."

LTL specifies properties of infinite-state trajectories. For a set AP of atomic propositions, an infinite sequence, $\sigma = \sigma(0)\sigma(1)\sigma(2)\ldots$, is defined by the values of the function $\sigma : \mathbb{N} \to 2^{\mathrm{AP}}$, which includes all propositions $p \in \mathrm{AP}$ that are true at time $t \in N$. The infinite sequence $\sigma^k = \sigma(k)\sigma(k+1)\sigma(k+2)\ldots$ is the part of the sequence σ starting at time k. The satisfaction relation $\models$ between trajectory σ and LTL formula φ is defined inductively as usual:

$\sigma \models p \Leftrightarrow \sigma(0)$ (i.e, $\sigma(0 \models p)$,

$\sigma \models \neg\varphi \Leftrightarrow \sigma \nvDash \varphi$,

$\sigma \models \varphi_1 \wedge \varphi_2 \Leftrightarrow \sigma \models \varphi_1$ and $\sigma \models \varphi_2$,

$\sigma \models X\varphi \Leftrightarrow \sigma^1 \models \varphi$,

$\sigma \models \varphi_1 \bigcup \varphi_2 \Leftrightarrow \bigvee k \geq 0 \; \sigma^k \models \varphi_2$ and $\bigwedge j < k \; \sigma^j \models \varphi_1$.

From a linguistic point of view, LTL specifies properties of the infinite sequence $\sigma = \sigma(0)\sigma(1)\sigma(2)\ldots$ with symbols $\sigma(i) \in 2^{\mathrm{AP}} = \Sigma$.

---

[39]Lennarston and Quin-Shan Jia (2020).

The infinite set $\{0, 1, 2, \ldots\}$ of natural numbers represents the smallest infinite ordinal number ω. Therefore, $\sigma \in \Sigma^\omega$ can be understood as infinite symbolic words. Subsets $\mathcal{L} \subseteq \Sigma^\omega$ are called ω-languages, which can also be used to specify properties equivalent to LTL specifications. In Section 3.3, it was explained that regular ω-languages with infinite words can be recognized by Büchi automata, while finite words of regular languages can be recognized by finite automata.

A Büchi automaton is defined as a tuple, $B = (\mathcal{Q}, \Sigma, \delta, q_0, \mathcal{Q}_m, q_f)$, with a finite set $\mathcal{Q}$ of states, a finite set $\Sigma \subseteq 2^{AP}$ of symbols, transition function, $\delta : \mathcal{Q} \times \Sigma \to 2^S$, initial state q_0, a set $\mathcal{Q}_m$ of marked goal states, and a forbidden state q_f. A Büchi automaton must reach a marked state infinitely mans times, while a finite automaton must reach it one time. In LTL, the acceptance condition of a finite automaton can be expressed as $\Diamond M$ with state label M for all marked states. The acceptance condition of Büchi automaton is described as $\Box \Diamond M$. In Fig. 4.6, a marked state is represented by a double circle. The forbidden state q_f is represented by a cross, which is reached for all non-accepted words.

In order to specify properties on state labels in temporal logic, an MDP is defined for RL. An MDP is a tuple $\mathcal{M} = (S, \mathcal{A}, P, s_0, AP, \lambda, \rho)$ with a countable set S of states, a finite set $\mathcal{A}$ of actions and set $\mathcal{A}(s)$ of available actions in state s, transition probability function $P : S \times \mathcal{A} \times S \to [0, 1]$ such that $P(s, a, s')$ is the transition probability for transition (s, a, s') from state s to state s' for action $a \in \mathcal{A}(s)$ with $\sum_{s' \in S} P(s, a, s') = 1$, initial state s_0, a set AP of atomic propositions, a state labeling function $\lambda : X \to 2^{AP}$, and a reward function $\rho : S \times \mathcal{A} \times S \to \mathbb{R}$ with the immediate reward $\rho(s, a, s')$ after transition (s, a, s'). The function value $\lambda(s)$ of the labeling function λ indicates those atomic propositions which are satisfied in state s.

In RL, an MDP should satisfy an LTL formula φ. Therefore, the corresponding Büchi automaton B_φ is synchronized with the MDP $\mathcal{M}$ by the definition of a product model $\mathcal{M} \otimes B_\varphi$.[40] For $M = (S, \mathcal{A}, P, s_0, AP, \lambda, \rho)$ and $B = (\mathcal{Q}, \Sigma, \delta, q_0, \mathcal{Q}_m, q_f)$, the synchronization is defined by

$$M \otimes B := (S \times \mathcal{Q}, \mathcal{A}, P_\otimes, (s_0, \delta(q_0, \lambda(s_0))), S \times \mathcal{Q}_m, S \times \{q_0\}, \rho_\otimes),$$

---

[40]Baier and Katoen (2008).

with transition probability

$$P_{\otimes}((s,q),a,(s',q')) = \begin{cases} P(s,a,s'), & \text{if } q' = \delta(q, \lambda s')), \\ 0, & \text{otherwise}, \end{cases}$$

and reward

$$\rho_{\otimes}(s,q) = \rho(s) + \begin{cases} \rho_M > 0, & \text{if } q \in \mathcal{Q}_m, \\ \rho_F < 0, & \text{if } q = q_f, \\ 0, & \text{otherwise}. \end{cases}$$

The state label $\lambda(s')$ of the target state s' in the MDP $\mathcal{M}$ should satisfy the related transition label in the Büchi automaton B. In this sense, a transition $((s,q),a,(s',q'))$ in the synchronized system with transition probability $P_{\otimes}$ is restricted by the condition $q' = \delta(q, \lambda(s'))$.

In Fig. 4.6, forbidden states are introduced for non-accepted words. The safety specification $\Box \neg q$ requires that no state with label q is accepted. In the Büchi automaton B_φ, the forbidden state 2 models that a transition of the MDP $\mathcal{M}$ to a state with label q is not accepted. In the reward function $\rho_{\otimes}$, a positive reward $\rho_M > 0$ is related to marked states and a negative reward $\rho_F < 0$ to forbidden states.

In Fig. 4.6, an MDP $\mathcal{M}$ has state labels p and q. For the specification $\varphi = \Diamond p \wedge \Box \neg q$, the corresponding Büchi automaton B_φ is shown with a forbidden state after transitions with label q. The synchronization $\mathcal{M} \otimes B_\varphi$ has two marked states because specification $\Diamond p$ is fulfilled in state 2 as well as 3 in $\mathcal{M}$.

In RL, learning is an optimization process which can be performed with, for example, a so-called Q-function. A Q-function is an action-value function $Q : S \times \mathcal{A} \to \mathbb{R}$ that depends on both the system state and possible actions. The function is estimated from system data, where the next state and resulting reward are considered as a result of a given action which is decided by the learning agent. The optimization of a control policy is performed by a stochastic procedure (e.g., dynamic programming) for the MDP. The policy is determined by the Q-function. It selects actions such that marked states, if possible, will be visited infinitely often. Forbidden states are avoided. In this way, an LTL formula is satisfied.

Markov decision process $\mathcal{M}$:

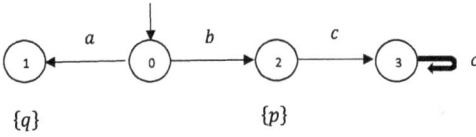

Büchi automaton B_φ for LTL specification $\varphi = \Diamond p \wedge \Box \neg q$:

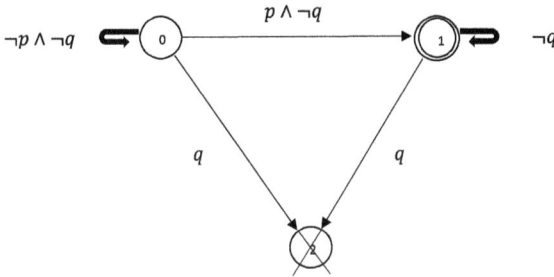

Synchronization $\mathcal{M} \otimes B_\varphi$:

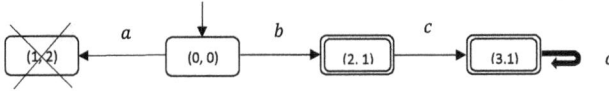

Fig. 4.6.   Synchronization $\mathcal{M} \otimes B_\varphi$ of a Markov decision process $\mathcal{M}$ and a Büchi automaton B_φ.[41]

4.2. Temporal Logic in Relativistic Physics

Traditional temporal logics assume classical physics and its concept of time as self-evident. But we have to consider different concepts of time in the different physical theories, such as classical physics, relativistic physics, thermodynamics, and quantum physics. In modern physics, it is well-known that the concepts of time, space, and matter change if research approaches the size of cosmic structures or elementary particles, the velocity of light, and the gravitation of black holes, i.e., in short, if measurements become more and more

---

[41]Lennarston and Quin-Shan Jia (2020, Fig. 1).

dependent on Planck's quantum constant and the constant of light velocity.

In special relativity, the mathematical structure of space-time is given by the four-dimensional Minkowskian geometry.[42] The four-dimensional points with three spatial and one temporal coordinate are physically understood as events. According to Einstein, an event y comes after an event x if a signal can be sent from x to y at most with the speed of light. Physically, the causal connections between events are defined in this way. Concerning temporal logic, the temporal operator $\Box$ of necessity relates to sentences on events which "will always be." It can be proven that modal sentences valid in this space-time structure are the theorems of the modal calculus S4.2. With respect to temporal logic, the modalities of this calculus relate to a branching-tree temporal logic. With respect to physics, the Minkowskian space-time of relativity theory is one of the most significant nonlinear-time structures.

4.2.1.   *Temporal logic of time frames*

The syntax of propositional modal logic consists of sentence letters $p, q, r, \ldots$, which are connected by Boolean connectives and the modal operator $\Box$ ("it is now and will always be the case that"). The modal operator $\Diamond$ ("it will be at some time that") is defined as usual with $\Box$ as $\neg\,\Box\,\neg$.[43]

The semantics of this formal language relates to time frames.[44] A time frame is defined as $\mathcal{T} = (T, \leq)$ with a nonempty set T of events and a reflexive and transitive ordering. A frame is called directed iff any two elements of T have an upper bound. A $\mathcal{T}$-valuation is a function V which assigns to each sentence letter p the set $V(p) \subseteq \mathcal{T}$ of events at which p is true. The valuation function can be extended by inductive definition to all sentences $A, B, C, \ldots$ defined with Boolean connectives. For modalities, it is defined as

$t \in V(\Box A)$ if and only if $t \leq s$ implies $s \in V(A)$,
$t \in V(\Diamond A)$ if and only if for some $s \in V(A)$, it is $t \leq s$.

---

[42]Mainzer (2010, Chapter III.1).
[43]Prior (1967).
[44]Goldblatt (1980).

The reflexivity of $\leq$ expresses the meaning of modal operator $\square$ ("it is and will always be"). A sentence A is valid in time frame $\mathcal{T} = (T, \leq)$ according to this definition:

$\mathcal{T} \models A$ iff $V(A) = T$ for every $\mathcal{T}$-valuation V.

Time frames may have an analogous structure, which is defined with a homomorphic mapping. Time frames $\mathcal{T} = (T, \leq)$ and $\mathcal{T}' = (T', \leq')$ are called p-homomorphic (abbreviation: $\mathcal{T} \twoheadrightarrow \mathcal{T}'$) iff there is a (surjective) function $f : T \to T'$ (p-homomorphism) with

(i) $t \leq s$ implies $f(t) \leq' f(s)$,
(ii) $f(t) \leq' v$ implies that there is some $s \in T$ with $t \leq s$ and $f(s) = v$.

For p-homomorphic time frames with $\mathcal{T} \twoheadrightarrow \mathcal{T}'$, it is $\mathcal{T} \models A$ only if $\mathcal{T}' \models A$ for any sentence A.

A subset $T' \subseteq T$ is called future-closed under $\leq$ if, for all $t \in T'$ with $t \leq s$, it is $s \in T'$. A time frame $T' = (T', \leq')$ with a future-closed subset $T' \subseteq T$ is called a subframe of time frame $\mathcal{T} = (T, \leq)$. Because of the transitivity of $\leq$, for each t, the set $\{s | t \leq s\}$ is the base of a subframe of $\mathcal{T}$ generated by t. An element $0 \in T$ is called an initial point of time frame $\mathcal{T}$ if $0 \leq s$ is for all $s \in T$. Therefore, t is an initial point of the subframe generated by t.

For subframe $\mathcal{T}'$ of $\mathcal{T}$, it is $\mathcal{T} \models A$ only if $\mathcal{T}' \models A$ for any sentence A.

The semantics of time frames can be related to the formal modal calculus S4.2, which is axiomatized by the following formulae:

(i) $\square(A \to B) \to (\square A \to \square B)$,
(ii) $\square A \to A$,
(iii) $\square A \to \square\square A$,
(iv) $\Diamond\square A \to \square\Diamond A$

and rules

(i) modus ponens: $\frac{A \quad A \to B}{B}$,
(ii) necessitation: $\frac{A}{\square A}$.

Formula (i) is valid on all frames because it expresses a general property of necessity independent of the properties of $\leq$. Formula (ii) is valid because of the reflexivity of $\leq$. Formula (iii) is valid because of the transitivity of $\leq$. Formula (iv) is valid if $\leq$ is directed. Therefore, the following version of soundness is true for the modal calculus S4.2 with respect to the semantics of time frames:

$\vdash_{S4.2} A$ implies that A is valid in all directed frames.

Completeness can also be proven for S4.2 in the following sense:

$\nvdash_{S4.2} A$ implies that there is a finite generated and directed time frame $\mathcal{T}$ with $\mathcal{T} \nvDash A$.

The equivalence relation on T with

$$t \approx s \quad \text{iff } t \leq s \quad \text{and} \quad s \leq t$$

leads to clusters $\bar{t}$ and $\bar{s}$ as equivalence classes. They can be ordered by an ordering

$$\bar{t} \leq \bar{s} \quad \text{iff } t \leq s,$$

which is antisymmetric because $\bar{s} \leq \bar{t}$ and $\bar{t} \leq \bar{s}$ implies $\bar{t} = \bar{s}$. Therefore, a time frame can be represented as a partially ordered collection of clusters.

A final element s in a time frame $\mathcal{T}$ with $t \leq s$ for all $t \in T$ is denoted by ∞. If time frame $\mathcal{T}$ is directed and finite, then it must have at least one final point. All final points are $\approx$-equivalent and belong to cluster $\bar{\infty}$. A time frame $\mathcal{T}$ can be extended by a unique final point $\infty \notin T$ with $\mathcal{T}^{\infty} := (T \cup \{\infty\}, \leq)$ and appropriately extended ordering $\leq$ of $\mathcal{T}$ for ∞. As the final point is an upper bound for any two elements, time frame $\mathcal{T}^{\infty}$ is directed.

In order to construct an appropriate time frame for the Minkowskian space-time of relativity, an infinite binary-branching frame $\mathcal{B} = (B, \leq)$ is introduced.[45] The elements $x \in B$ are finite sequences $x = x_1 x_2, \ldots, x_n$, with $x_i \in \{0, 1\}$ and length $l(x) = n$. The empty sequence $x = \emptyset$ with $l(x) = 0$ is included.

The ordering for sequences $x = x_1 x_2, \ldots, x_n$ and $y = y_1 y_2, \ldots, y_n$ is defined by

$x \leq y$ iff x is an initial element of y,

Iff $n \leq m$, and $y = x_1 x_2, \ldots, x_n y_{n+1} y_{n+2}, \ldots, y_m$.

---

[45]Goldblatt (1980).

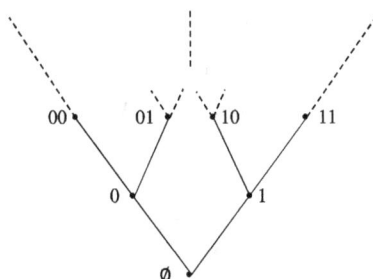

Fig. 4.7. Infinite binary-branching frame $\mathcal{B} = (B, \leq)$.

Obviously, the infinite binary-branching frame is partially ordered with $\emptyset$ as the initial point. The successors of x in $\mathcal{B}$ are the sequences that extend x. There are exactly two immediate successors 0 and 1 of x. In Fig. 4.7, the tree of the infinite binary-branching frame $\mathcal{B} = (B, \leq)$ is illustrated. The length $l(x)$ can be considered as the level of x in $\mathcal{B}$. In Section 1.2, this tree structure was already introduced as a binary fan of intuitionistic mathematics.

It can be proven that the infinite binary-branching frame $\mathcal{B}$ is p-homomorphic to any finite generated time frame.

Modal logic S4 contains the axioms of S4.2 without axiom (iv). Any non-theorem for S4 is falsifiable on a finite generated reflexive and transitive frame. The infinite binary-branching frame $\mathcal{B}$ is p-homomorphic to any finite generated time frame. For p-homomorphic time frames with $\mathcal{T} \twoheadrightarrow \mathcal{T}'$, it is $\mathcal{T} \models A$ only if $\mathcal{T}' \models A$ for any sentence A. Therefore, for any sentence A, it holds that

$$\vdash_{S4} A \quad \text{if and only if} \quad \mathcal{B} \models A.^{46}$$

A characteristic frame for the calculus S4.2 can be found by extension of $\mathcal{B}$ with an infinite final cluster at the top of $\mathcal{B}$. For an infinite set $\Omega = \{\infty_0, \infty_1, \ldots, \infty_n, \ldots\}$ of elements disjoint from B, a frame $\mathcal{B}^\Omega = (B \cup \Omega, \leq)$ is defined with an extended ordering $\leq$ with respect to the elements of Ω.

It can be proven that $\mathcal{B}^\Omega \twoheadrightarrow \mathcal{T}$ (i.e., $\mathcal{B}^\Omega$ is p-homomorphic to $\mathcal{T}$) for any finite directed and generated frame $\mathcal{T}$. It follows that

$$\vdash_{S4.2} A \quad \text{if and only if} \quad \mathcal{B}^\Omega \models A \quad \text{for any } A.^{46}$$

---

[46]Goldblatt (1980, p. 225).

4.2.2.   Time frames of Minkowskian space-time

The metric of an n-dimensional Minkowskian geometry is defined by $\mu(x) := x_1^2 + x_2^2 + \cdots + x_n^2$ with n-tupel $x = (x_1, x_2, \ldots, x_n)$ of real numbers. In special relativity, it is $n = 4$, with spatial coordinates x_1, x_2, x_3 and time coordinate $x_4 = t$ which is distinguished by the minus-sign in the Minkowskian metric.[47]

Mathematically, an n-dimensional space-time can be defined as a frame $\mathcal{T}^n = (\mathbb{R}^n, \leq)$ with

$$x \leq y \quad \text{iff} \quad \mu(y - x) \leq 0 \quad \text{and} \quad x_n \leq y_n \quad \text{for } x, y \in \mathbb{R}^n$$

$$\text{iff} \quad \sum_{i=1}^{n-1} (y_i - x_i)^2 \leq (y_n - x_n)^2 \quad \text{and} \quad x_n \leq y_n \quad \text{for } x, y \in \mathbb{R}^n.$$

$\mathcal{T}^n$ is a partially ordered and directed frame. An upper bound of x and y is $z := (x_1, x_2, \ldots, x_{n-1}, z_n)$ with $z^n = \sum_{i=1}^{n-1} (y_i - x_i)^2 + |x_n| + |y_n|$.

It can be proven that $\mathcal{T}^n$ and $\mathcal{T}^{n+1}$ are p-homomorphic (i.e., $\mathcal{T}^n \twoheadrightarrow \mathcal{T}^{n+1}$).[48]

Physically, Minkowskian space-time is $\mathcal{T}^4$. Elements x and y are called "events." The relation $x \leq y$ means that a signal can be sent from event x to event y with a speed at most that of light in the causal future of x. The spatial and the time coordinates can be chosen such that the speed of light is one unit of distance per unit of time. Then, in the frame $\mathcal{T}^2$, the future for each event $e = (x, y)$ in the plane consists of all events on or above upwardly directed rays of slopes $+1$ and -1 from e (Fig. 4.8).[49]

It can be proven that for any sentence A, it holds that

$$\vdash_{S4.2} A \text{ iff } \mathcal{T}^n \models A \text{ iff } I \models A \text{ for the unit box } I := [0,1) \times [0,1).$$

In the time frame $\mathcal{T}^n$, a relation of events can be defined with

$$x \prec y \quad \text{iff} \quad \mu(y - x) \leq 0 \quad \text{and} \quad x_n \leq y_n.^{50}$$

---

[47]Mainzer (1988, pp. 39–44, Chapter 3.3: Minkowski's spacetime).
[48]Goldblatt (1980, pp. 225–227).
[49]Goldblatt (1980, p. 227, Fig. 3).
[50]Goldblatt (1980, pp. 232–234).

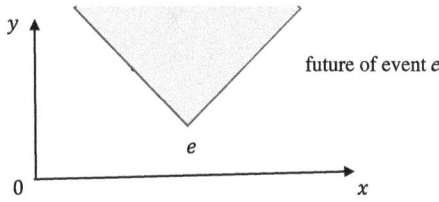

Fig. 4.8. Future in Minkowskian space-time $\mathcal{T}^2$.

Physically, $x \prec y$ means that a signal can be sent from event x to event y at less than the speed of light. For the reflexive relation

$$xRy \quad \text{iff} \quad x = y \text{ or } x \prec y,$$

the valid sentences on time frame $(\mathcal{T}^n, R)$ are exactly the theorems of the logical calculus S4.2.

The standard models of relativistic cosmology allow a mathematical solution that cosmic expansion will lead to a contraction of the universe, with a final collapse to a singularity. In this case, any future path in space-time will end in the singularity. Mathematically, in a time frame, the existence of a singularity means

$$\bigwedge x \bigvee y (x \le y \wedge \bigwedge w(y \le w \to y = w)).$$

In a directed partially ordered frame, there can be only one unique final event y with the singularity condition. The reason is that if y has no successors, then an upper bound for y and the other point can only be y itself.

The calculus K2 extends the calculus S4.2 by the axiom

$$\Box \Diamond A \to \Diamond \Box A.$$

This formula is valid on frames with the singularity condition.

In temporal logic, we must distinguish two irreflexive orderings with different consequences for space-time. There is the already defined ordering

$$x \prec y \quad \text{iff} \quad \mu(y - x) \le 0 \quad \text{and} \quad x_n \le y_n,$$

which must be distinguished from the after-relation in temporal logic

$$xay \quad \text{iff} \quad x \ne y \quad \text{and} \quad x \le y.$$

The difference between both orderings becomes obvious in terms of the validity of modal sentences. Concerning the after-relation a,

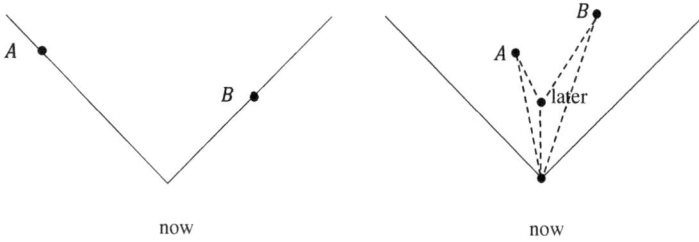

two propositions A and B are considered, which hold in the future at two points that can only be reached from now by traveling in opposite directions at the speed of light (Fig. 4.9(left)). In this case, $\Diamond A \vee \Diamond B$ will be true now but never again. It follows that

$$\Diamond A \wedge \Diamond B \rightarrow \Diamond(\Diamond A \wedge \Diamond B)$$

is not valid for the past-relation a.

For the $\prec$-relation, the formula is true. The reason is that a slower-than-light travel can always be realized with faster velocity. Therefore, the traveler can wait for some time and travel to A and B at a greater speed (Fig. 4.9(right)). This difference between the two orderings is true for all time frames $\mathcal{T}^n$ with $n \geq 2$.

It is remarkable that the validity of modal sentences is dimension dependent in the Minkowskian geometry of relativity theory. In three-dimensional space-time $\mathcal{T}^3$, there are at least three points that can only reached by traveling in different directions with the speed of light. In Fig. 4.10, the future of an event e is illustrated by the well-known Minkowskian right circular cone, which is centered in e.[52] In $(\mathcal{T}^3, a)$ (also in $(\mathcal{T}^n, a)$ with $n \geq 3$), the following modal sentence can be falsified for $ij \in \{1, 2, 3\}$:

$$\left(\bigwedge_i \Diamond p_i \right) \wedge \left(\bigwedge_{i \neq j} \Box(p_i \rightarrow \neg \Diamond p_j) \right) \rightarrow \bigvee_{i \neq j} (\Diamond(\Diamond p_i \wedge \Diamond p_j)).$$

But this sentence is true in $(\mathcal{T}^n, \prec)$ with $n \geq 2$ and in $(\mathcal{T}^2, a)$. The two-dimensional Minkowskian space-time consists of one space

---

[51]Goldblatt (1980, p. 234).
[52]Mainzer (1988, pp. 40–42).

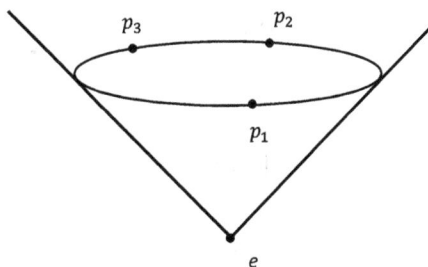

Fig. 4.10. Minkowskian future cone in $\mathcal{T}^3$.

dimension and one time dimension. In the corresponding temporal logic, there is an algorithm to determine whether a given temporal formula is valid over the two-dimensional Minkowskian space-time or not. For higher dimensions, such an algorithm of decidability is not known.

4.2.3. Modal and temporal logics for continuous and discrete space-time

In the studies of Minkowskian space-time, one can distinguish the languages of modal and temporal logic. The language of modal logic is the language $\mathcal{L}$ of propositional logic extended by the modal operator $\Box$ of necessity. The language of temporal logic consists of the language $\mathcal{L}$ of propositional logic extended by the temporal operators G ("necessarily in the future") and H ("necessarily in the past"). There are dual operators with the following definitions[53]:

$$\Diamond := \neg\Box\neg \ (\text{"possibly"}),$$
$$F := \neg G\neg \ (\text{"possibly in the future"}),$$
$$P := \neg H\neg \ (\text{"possibly in the past"}).$$

In general, the semantics of these formal languages can be defined by Kripke models. A Kripke model is a structure $\mathcal{M} := (W, R, V)$, with set W of points ("worlds" or "times"), binary relation R on W, and a function V from the set of propositional variables to the set 2^W of all subsets of W. For a propositional variable p the set

---

[53] Uckelman and Uckelman (2007).

$V(p)$ contains all worlds or times when the proposition p is true. The semantic truth in a (Kripke) model is defined recursively for the Boolean connectives $\neg, \wedge, \vee$, and $\rightarrow$ in the usual way. For the modal and temporal operators, the necessary truth in the future and in the past of a formula φ in a world or time x of a Kripke model $\mathcal{M}$ is defined by

$$\mathcal{M}, x \models \Box\varphi \text{ iff } \mathcal{M}, y \models \varphi \quad \text{for all } y \text{ with } xRy,$$
$$\mathcal{M}, x \models G\varphi \text{ iff } \mathcal{M}, y \models \varphi \quad \text{for all } y \text{ with } xRy,$$
$$\mathcal{M}, x \models H\varphi \text{ iff } \mathcal{M}, y \models \varphi \quad \text{for all } y \text{ with } yRx.$$

The modal logic K consists of all axioms of propositional logic and the modal axiom

$$\Box(p \rightarrow q) \rightarrow (\Box p \rightarrow \Box q)$$

and the rules of modus ponens and inference of $\vdash \Box\varphi$ from $\vdash \varphi$.

The corresponding temporal logic K_t consists of all axioms of propositional logic extended by axioms of the temporal operators G and H with

(i) $H(p \rightarrow q) \rightarrow (Hp \rightarrow Hq)$,
(ii) $G(p \rightarrow q) \rightarrow (Gp \rightarrow Gq)$,
(iii) $p \rightarrow HFp$,
(iv) $p \rightarrow GPp$

and the rules of modus ponens and inference of $\vdash H\varphi$ and $\vdash G\varphi$ from $\vdash \varphi$.

One can restrict to the modal language with the constraint that the relation R must be reflexive and a proposition p true, resp. possible, at a point whenever p is necessary, resp. true, at that point. One can also restrict to the temporal language with the constraint that $\Box p$ (necessarily p) is a definition for $p \wedge Gp$ (p and p always true in the future).

Mathematically, Kripke models can be considered for continuous worlds $W = \mathbb{R}^n$ and discrete worlds $W = \mathbb{Z}^n$. Continuous worlds are interesting for the physical application of the Minkowskian space-time $(\mathbb{R}^4, \preceq)$, with $\preceq$ as relation R of Kripke models. The set $\{y | x \preceq y\}$ is the Minkowskian future light cone of event x. This is the set of all future physical events which are accessible from x. Events outside the future light cone are not accessible from x and therefore causally undetermined. Mathematically, $(\mathbb{R}^n, \preceq)$ is isomorphic to $(\mathbb{R}^n, \leq)$, which is the rotation of $(\mathbb{R}^n, \preceq)$ with an angle of $45°$ in Fig. 4.8 (for $n = 2$). Relation $\preceq$ is reflexive. There are two irreflexive orderings $\prec$ and a in modal and temporal logic, which were already studied in the previous section with their different consequences.

A modal logic of the frames $(\mathbb{R}^n, \leq)$ with $n \geq 2$ is S4.2, which consists of calculus K and the modal axioms

(i) $\Box p \to \Box\Box p$,
(ii) $\Box p \to p$,
(iii) $\Diamond\Box p \to \Box\Diamond p$.

S4.2 is finitely axiomatizable and complete.[54]

The modal logic of the frames $(\mathbb{R}^n, \prec)$ $n \geq 2$ is L_2, which consists of K and the modal axioms

(i) $\Diamond\Diamond p \to \Diamond p$,
(ii) $\Diamond(p \vee \neg p)$,
(iii) $(\Diamond p \wedge \Diamond q) \to \Diamond(\Diamond p \wedge \Diamond q)$,
(iv) $\Diamond\Box p \to \Box\Diamond p$.[55]

Axiom (ii) expresses that there are no dead ends in the frames, and every n-tupel can access another n-tupel. Axiom (iii) means a version of density in the frames: For all x, y_1 and y_2, there is a z such that if $x \prec y_1$ and $x \prec y_2$, then $x \prec z$, $z \prec y_1$, and $z \prec y_2$.

It is remarkable that the modal logics of the reflexive and irreflexive versions of frames $\mathbb{R}^n$ are the same for all dimensions $n > 1$. Concerning the temporal after-operator a, no axiomatization of modal or temporal logics is known for frames $(\mathbb{R}^n, a)$ for all n. It is only known that the temporal logics for $(\mathbb{R}^n, a)$ must be distinct for all n. Whereas the modal logics for $(\mathbb{R}^n, \leq)$ is the same for all $n \geq 2$, the modal logics for $(\mathbb{Z}^n, \leq)$ and $(\mathbb{Z}^n, a)$ are distinct for all n.[56] Obviously, discrete structures seem to be much more complicated than continuous ones.

---

[54]Goldblatt (1980, pp. 220–221).
[55]Shapirovsky and Shehtmann (2003, pp. 437–459).
[56]Phillips (1998, pp. 545–553).

4.2.4.   *Conceptual foundations of general relativity*[57]

A main requirement of this book is that temporal logic must consider the change in the time concepts of physics from Aristotle and Newton to Einstein and the quantum world. The axioms of a calculus in temporal logic depends on an intuitive understanding of conceptual foundations of time. Therefore, before analyzing formal systems of temporal logic in general relativity, we start with a short informal reminder on the conceptual foundations of general relativity. In special relativity theory, an absolute space-time frame is abandoned, and all measurements and observations are taken relative to localized frames of reference under the assumption of the velocity of light as maximum speed. But accelerations according to Newton's law of gravitation are not considered. In general relativity theory, special relativity is extended, and a relativistic gravitational law included.

When Einstein extended his investigation of space-time to accelerated reference frames in 1907, he assumed that the acceleration effect of a reference frame cannot be distinguished from the effect of a gravitational field. His thought experiment of an observer who is in a closed box without contact with the outside world and can only observe the movements of bodies in this box is well known. All masses experience the same constant acceleration downward in a homogeneous gravitational field. Also, with respect to a box moving upward with the same acceleration, free mass points of any mass experience this acceleration downward. An observer in the box can therefore not decide by measurement whether the box is constantly accelerated or whether it is in a homogeneous gravitational field.

Equivalently, one can also say that the effects of a homogeneous gravitational field on masses can be simulated in an indistinguishable way from those in reference to a suitably accelerated reference frame. In other words, by referring to a freely falling reference frame (e.g., an elevator), all effects of a homogeneous gravitational field can be eliminated, i.e., weightlessness prevails.

So far, only the special case of homogeneous gravitational fields has been considered. In an inhomogeneous gravitational field with

---

[57]Mainzer (2010, Chapter III.2).

different gravitational effects, however, sufficiently small areas can always be considered, in which gravitational effects hardly change and whose effects can therefore be approximated by a homogeneous gravitational field.

The equivalence principle therefore states that at least locally, i.e., in very small space-time sections in which the gravitational field does not change, an inertial system can be chosen, whereby the gravitational effect is canceled. Locally, therefore, the laws of special relativity apply without gravity. Einstein's thought experiment is realized today by passengers in an airplane, who experience weightlessness during free fall in the Earth's gravitational field (e.g., nosedive).

Even with frequency measurements on light beams, an observer cannot distinguish between a constantly accelerated reference frame and a homogeneous gravitational field. Like a thrown stone, light loses its energy when it rises in the gravitational field against the effect of a gravitational attraction. Its frequency decreases, and its color shifts toward the red part of the spectrum, i.e., toward longer wavelengths. If a beam of light is sent from a wall to the opposite wall in a closed box at right angles to the direction of acceleration or gravitational effect, it is indistinguishably curved toward the ground in both cases. This is the case observed by Eddington in 1917 with the deflection of light rays in the gravitational field of the sun.

According to the equivalence principle, there are again local effects of gravitational fields that are globally inhomogeneous. Freely falling bodies in inhomogeneous gravitational fields exhibit relative accelerations among themselves. In the absence of gravity, free point particles and light rays move with constant speed on straight lines, i.e., without relative acceleration among themselves. If a gravitational field is switched on, for example, by bringing a large mass into the vicinity, these paths are bent or exhibit relative accelerations. Gravitational effects thus correspond geometrically to orbital curvature. In this sense, an inhomogeneous gravitational field corresponds globally to a curved space-time.

At first, this is reminiscent of a well-known case of the spherical geometry on the Earth's sphere, as it is known in geodesy. The straightest paths on the curved surface of the Earth (e.g., the shortest connection between two places) are circular arcs. If one chooses arbitrarily small sections, one obtains approximate distances: An arc is mathematically composed of infinitesimally small straight lines. However, the geometry of straight lines is Euclidean. Therefore, in Riemannian differential geometry of curved spaces (e.g., spherical geometry on the surface of a sphere), it is said that Euclidean geometry applies locally in infinitesimally small areas (i.e., approximately also in the environment of an inhabitant of the Earth).

Similarly, Einstein's equivalence principle of the general theory of relativity with curved space-time demands that the uncurved (flat) space-time of Minkowski geometry applies locally in infinitesimally small areas, i.e., the physical laws are locally Lorentz-invariant as in the special theory of relativity. In general, the physical laws are covariant, i.e., they retain their shape (shape invariance) in general (also curved) coordinate transformations.

The local Lorentz invariance determines the time and causality relations in a general relativistic gravitational field: If a point (event) can be connected to another point (event) by a time-like curve, then a signal can be sent from this point to the other point but not vice versa. In fact, Einstein's gravitational equation also allows for solutions in which time-like curves describe closed arcs. In such a world, the physical paradox of an astronaut traveling into his own past would occur. This case raises questions about relativistic cosmology.

Empirical confirmation of Einstein's gravitational equation was provided by light deflection, propagation delay, and perihelion rotation. For time logic, time dilation due to gravitation is of great interest. Clocks that are closer to the center of the earth and thus deeper in the Earth's gravitational field run, albeit minimally, but measurably slower.

This gravitational time dilation, according to general relativity, can be illustrated by a thought experiment with twins. A twin who has been exposed to strong gravitation on the surface of a very dense celestial body (e.g., neutron star) will be significantly younger than his twin brother when he returns to Earth. This gravitational effect on time must be distinguished from time dilation in special relativity, which depends on relativistic kinematic effects.

Different developmental models of the universe can be derived on the basis of the general theory of relativity.[58] The first empirical evidence for a temporal evolution of the universe goes back to

---

[58]Weinberg (1972), Mainzer (2010, Chapter III.3).

the astronomer Hubble. In 1929, he discovered that the speed of the galaxies' escape motion increases with the distance between a galaxy cluster and its observer. This is how he interpreted the observation: The light of very distant galaxies shifts to the red region of the spectrum, i.e., to longer wavelengths. The basis of this explanation is the optical Doppler effect, according to which the wavelengths of light emitted by a moving light source appear longer to an observer at rest as the light source moves away and shorter as it approaches.

The cosmological principle assumes that all points in space physically undergo the same evolution correlated in time so that all points at a fixed distance appear to an observer to be just at the same stage of evolution. In this sense, the spatial state of the universe must appear homogeneous and isotropic to the observer at all times in the future and past.

The cosmological principle is based on the (simplified) assumption that matter is distributed uniformly (homogeneously) on average in the universe and that its properties remain the same, regardless of an observer's viewing direction (isotropy). In fact, homogeneity and isotropy are statistically confirmed, at least approximately, by observing the distribution of stars from Earth.

Assuming the cosmological principle, Einstein's gravitational equation gives rise to Friedmann's standard models of cosmic evolution. Mathematically, the three cases of positive, Euclidean (flat), and negative curvature can be distinguished for homogeneous and isotropic spaces. Flat curvature means no or zero curvature. Cosmic evolution in the three standard models is described by the development of the time-dependent "world radius" $R(t)$ and an energy density function in a first-order differential equation that can be derived from Einstein's gravitational equation.

According to the singularity theorems of Penrose[59] and Hawking[60] (1970), it also follows from general relativity that the standard cosmic models must have an initial space-time singularity with infinite curvature. Cosmologically, it is interpreted as the "big bang" of the universe. After that, the universe initially expands very rapidly (inflationary universe), then slows down to the velocity of expansion which

---

[59]Penrose (1965).
[60]Hawking and Penrose (1970).

has been observed by astronomers. In the standard model with positive curvature, the expansion reverses to a collapse, which represents a final singularity. One then speaks of a closed universe. Analogously, in the two-dimensional case, a spherical surface on a sphere with positive curvature is also "closed" and finite. For the other two standard models with flat and negative curvatures, the expansion continues indefinitely with greater or lesser speed. One then also speaks of open (potentially infinite) universes. In the two-dimensional case, the flat curvature corresponds to an unlimited Euclidean plane, while the negative curvature corresponds to an unlimited saddle-like surface.

The singularity theorems also predict the possibility of very small regions of relativistic space-time, where space-time can become extremely curved and, therefore, gravity can become maximally large. Astrophysically, these singularities are interpreted as "black holes" preceded, for example, by the death of a large star through gravitational collapse. For this purpose, a three-dimensional spatio-temporal surface ("absolute event horizon") is assumed, which "swallows up" all incoming signals from outside and does not allow any signals or particles to escape. In the center of this absolute event horizon, the spatio-temporal singularity is assumed, in which the curvature of space-time becomes infinite. It is therefore an absolute end point for causal time signals.

A thought experiment with an astronaut approaching a black hole explains the consequences. If this astronaut passes the (relative) event horizon of the black hole and from then on sends a light signal at equal intervals according to his clock to a space station outside the event horizon, the light signals will be received there according to the space station's clock at ever greater intervals and with a red shift to ever longer wavelength light until the astronaut's time comes to a standstill as seen from outside, and the light signals can no longer be received because of infinite curvature and maximum gravity.

Finally, other cosmological principles have been proposed. The partial cosmological principle, according to Gödel,[61] is interesting for the discussion of time. According to this principle, the universe is only homogeneous but not isotropic. This solution also allows for Einstein's gravitational equation. The cosmic space-time associated with it has

---

[61]Gödel (1949).

the possibility of closed, time-like world lines, which allow for high-grade science fiction situations. In Gödelian space-time, an observer could embark on a journey into the past, only to encounter his or her own former self. However, the microwave background radiation as a relic from the primeval times of the universe empirically refutes this model since it is not only highly homogeneous but also isotropic.

But if the cosmological principle is correct, then the question arises as to how the initial singularity of time and the high-degree symmetry of the universe with homogeneity and isotropy are to be explained physically. Obviously, the theory of relativity is no longer sufficient for this. Modern cosmology rather merges with quantum physics and elementary particle physics to form a research program in which the temporal evolution of the universe is to be explained.

4.2.5. *Temporal logics of general relativity*

In the sixth task of his famous list of open mathematical problems (1900), Hilbert requests an axiomatization of physics. Historically, he suggested an axiomatization of general relativity in 1915, which was (according to Einstein's critic) physically false. An axiomatization of special relativity can be introduced in first-order logic (FOL). This formal language distinguishes two sorts of objects: There is a sort Q of quantities (e.g., real numbers) with symbols $+$ and $*$ for operations and $<$ for a 2-ary predicate. The other sort B concerns physical "bodies" (e.g., masses, waves). There are special bodies, such as observers Ob and photons Ph (as elements of light), which are formally considered as 1-ary predicate symbols of sort B. An elementary proposition "observer o observes body b at space-time location p" has the predicative form $W(o, b, p)$.

A model of this language is defined by $\mathfrak{M} := (Q, +, *, <; B, Ob, Ph; W)$, with an ordered field $(Q, +, *, <)$ (e.g., real numbers), subsets $Ob \subseteq B$, $Ph \subseteq B$, and relation $W \subseteq B \times B \times Q^n$ (e.g., $Q = \mathbb{R}$ and $p = (p_1, \ldots, p_n) \in \mathbb{R}^n$). Without going into the details, physical concepts of relativity can be translated into the terms of FOL.[62] For example, a straight line from $p \in Q^n$ to $q \in Q^n$ with $p \neq q$ is defined as the set $\{p + x * (p - q) | x \in Q\}$. In this case, the symbols $+, -,$ and $*$ denote the usual operations of a vector space. The world

---

[62] Andréka *et al.* (2006).

line of a body b as observed by an observer o is defined as the set of space-time locations where o observes b to be present, i.e.,

$$w\text{line}_o(b) := \{p \in Q^n | W(o, b, p)\}.$$

The spatial distance and time distance between $p, q \in Q^n$ is defined by

$$\text{space}(p, q) := \sqrt{(p_2 - q_2)^2 + \cdots + (p_n - q_n)^2} \text{ and}$$

$$\text{time}(p, q) := |p_1 - q_1| = \sqrt{(p_1 - q_1)^2}.$$

Then, the speed to reach q from p can be defined by

$$\text{speed}(p, q) := \frac{\text{space}(p, q)}{\text{time}(p, q)}, \text{ when time } (p, q) \neq 0.$$

An essential axiom of special relativity is the demand that the speed of light is finite (nonzero) and isotropic (direction independent). This so-called light axiom is formalized in an FOL formula:

$$\bigwedge_{o \in \text{Ob}} \bigwedge_{ph \in \text{Ph}} ((w\text{line}_o(ph) \text{ is a straight line}) \wedge$$

$$\bigwedge_{p,q \in Q^n} (p \neq q \rightarrow (\text{speed}(p, q) = 1 \text{ iff}$$

$$\bigvee_{ph \in \text{Ph}} W(o, ph, p) \wedge W(o, ph, q)))).$$

Roughly, the theory of special relativity consists of Newtonian mechanics without the assumption of absolute time but extended by the light axiom. There is a formal system of axioms with special relativity as a model. In the sense of Tarski semantics, a formula φ of this system is a semantic implication of a formal system Ax of axioms (Ax $\models \varphi$) iff for every possible model $\mathfrak{M}$, it holds that if $\mathfrak{M} \models$ Ax, then $\mathfrak{M} \models \varphi$. In the Tarskian sense, the suggested formal system of special relativity is consistent.

As mentioned in the preceding section, in general relativity, gravitation can be studied by acceleration effects. Therefore, in a formal system of general relativity, new axioms are introduced for accelerated observers as a new sort of objects besides inertial observers who move with inertial systems on straight lines. General relativity is locally special relativity. Thus, a key axiom of accelerated observers demands that at each moment, each accelerated observer sees the nearby world for a short moment as an inertial observer does. The FOL axiom system of general relativity has the (smooth) Lorentz manifolds over ordered real, closed fields as models.[63]

4.2.6. *Temporal logic of black holes*

From a logical point of view, it is a main difference between the axiom systems of special and general relativity that the special theory has only one model of Minkowskian space-time, while the general theory has many non-isomorphic ones. Cosmologically exotic models are black holes. Since their early mathematical derivation from the Einsteinian theory, there are nowadays strong empirical evidence of their existence in the universe. There are different mathematical models of black holes which are allowed by the Einsteinian theory. The simplest model is the Schwarzschild black hole because all masses are assumed to be concentrated in one point. After an observer falls into a Schwarzschild black hole, he or she can only survive for a finite time before being crushed at the singularity point. A more friendly type are the slowly rotating black holes. Their space-time is mathematically defined by the Kerr-metric.[64] There are forever rotating "world lines," where observers could survive. They could have exciting consequences for temporal logic and decidability problems in theoretical computer science, which are now discussed.

In more detail, a Lorentzian manifold of general relativity (M, g) consists of a smooth, oriented, and time-oriented four-manifold M with a smooth Lorentzian metric g. The Lorentzian metric is a solution to the Einsteinian equations with respect to a smooth stress–energy tensor T on M. The length of an at least continuously differentiable time-like curve $\gamma : \mathbb{R} \to M$ is defined by

$$\|\gamma\| = \int_{\mathbb{R}} \sqrt{-g(\dot{\gamma}(\tau), \dot{\gamma}(\tau))} d\tau,$$

---

[63] Andréka *et al.*, pp. 82–83.
[64] Hawking and Ellis (1973).

which is physically understood as a world line of an observer moving in (M, g). The image $\gamma(\mathbb{R})$ is the class of events in M met by the observer. The length $\|\gamma\|$ of the world line is the proper time measured by the observer, which can be finite or infinite. The causal past of an event $q \in M$ contains the events from which an observer can travel to q without exceeding the speed of light, i.e., formally,

$$J^-(q) = \{x \in M| \text{ there is a future-directed non-space-like}$$
$$\text{continuous curve joining } x \text{ with } q\}.$$

The causal future $J^+(q)$ of event q can be defined similarly.

With respect to temporal logic, there is a remarkable space-time with a world line of infinite proper time. Formally, a space-time (M, g) is called Malament–Hogarth if there is a future-directed time-like half curve $\gamma_P : \mathbb{R}^+ \to M$, with $\|\gamma_P\| = \infty$ and a point $p \in M$ ("Malament–Hogarth event") with $\gamma_P(\mathbb{R}) \subset J^-(p)$.[65]

In a Malament–Hogarth space-time, there is a future-oriented time-like curve $\gamma_O : [a, b] \to M$ from a point $q \in J^-(p)$ in the causal past of p to p with $\|\gamma_O\| < \infty$. The point q can be chosen in the causal future of the past end point of γ_P.

The existence of world lines γ_P and γ_O in a Malament–Hogarth space-time could be used for dramatic applications of computability and decidability problems[66]: A physical computer P as a technical realization of a Turing machine can be imagined to move along the curve γ_P of infinite proper time (Fig. 4.11). In a Malament–Hogarth space-time, this physical computer can perform arbitrarily long calculations. But in this space-time, an observer can also be assumed to be following the curve γ_O with finite proper time to meet the Malament–Hogarth event $p \in M$ in finite proper time. By the definition of a Malament–Hogarth event, it is $\gamma_P(\mathbb{R}) \subset J^-(p)$. Therefore, in p, the observer can get a decision after an arbitrarily long computation of the physical computer on world line γ_P because it can send a signal to the observer on γ_O at arbitrarily late proper time.

---

[65]Hogarth (1992).
[66]Hogarth (1994), Etesi and Németi (2002).

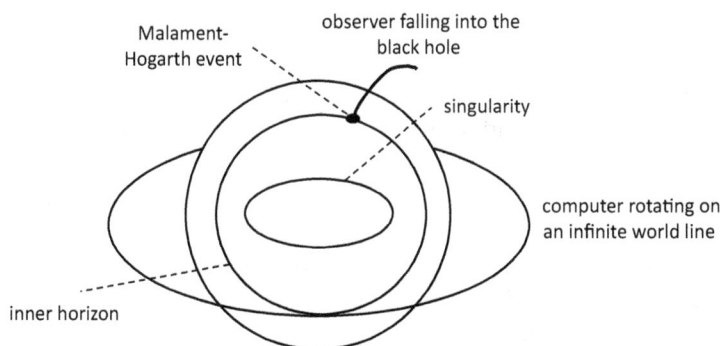

Fig. 4.11.   Computers overcome limits of computation in the Malament–Hogarth space-time of a rotating black hole.[67]

A thought experiment illustrates how the temporal operators "before" and "after" are concerned by the existence of a Malament–Hogarth event in a physical space-time. If the computer on γ_P is asked to check, for example, all theorems of set theory (ZFC) for consistency, then this task can be realized because the world line γ_P has infinite proper time. In the case of a contradiction, the computer sends a signal to the observer on world line γ_O. Now, the time of the signal's reception is decisive with respect to the Malament–Hogarth event: If the observer on γ_O receives a signal from γ_P before the Malament–Hogarth event p, then he or she can conclude that set theory is not consistent. If the observer does not get a signal before p, then after p, he or she knows that set theory is consistent. Starting the thought experiment in $\gamma_O(a) = q$ and ending it in the Malament–Hogarth event $\gamma_O(b) = p$, the observer on γ_O has proper time to decide a question which is not decided in mathematics.

Is a Malament–Hogarth space-time only a mathematical solution to the Einsteinian equations, such as a Gödel universe with travels in the past of the observer, but with less physical reality?[68] Actually, there are physical hints that make Kerr black holes with Malament–Hogarth space-time physically reasonable. The black hole in the center of the Milky Way is a possible candidate. In this case,

---

[67]Etesi and Németi (2002).
[68]Etesi and Németi (2002), Melia (2000).

a computer could be assumed to overcome limits of computability and decidability of classical Turing machines. In mathematical theory of computational complexity, there are hierarchies of degrees of computability and decidability.[69] In science fiction, one could imagine a future generation of mankind which colonizes the Milky Way and stations its computers on an infinite world line rotating about a Kerr black hole. Anyway, this thought experiment underlines that temporal logic cannot only be based on concepts of time in the tradition of Aristotle and Newton. The situation is quite similar to the temporal logic of quantum physics, which will be considered in the following sections. There are also problems which would be solvable if a physical quantum computer is realized (e.g., Shor algorithm in quantum cryptography). But, of course, a quantum computer will be realized much earlier than a computer "driven" by a black hole.

4.3.   Temporal Logic in Quantum Computing

Quantum computing is one of the most promising future technologies of computing. It also deeply rooted in epistemic and foundational concepts of quantum physics. Therefore, the basics of quantum computing are explained conceptually before the question of temporal logic in quantum computing is considered.[70]

4.3.1.   *Basics of quantum computing*

In a classical computer, a bit is either 0 or 1. A quantum bit of a quantum computer can be 0 as well as 1. Physically, a quantum system (e.g., a photon) can be vertically and, simultaneously, horizontally polarized. These alternative states are represented by the vectors $|0\rangle$ and $|1\rangle$ in two-dimensional Hilbert space $\mathcal{H}$ over the complex numbers. The non-superposed states $|0\rangle$ and $|1\rangle$ can be understood as basic elements of the two-dimensional vector space $\mathcal{H}$, i.e., $|0\rangle = \begin{pmatrix} 1 \\ 0 \end{pmatrix}$ and $|1\rangle = \begin{pmatrix} 0 \\ 1 \end{pmatrix}$.

---

[69]Mainzer (2018, Chapters 3 (Hierarchies of Computability), 11 (Complexity Theory of Real Computing), 15 (Digital and Real Physics)).
[70]Mainzer (2020).

According to the superposition principle of quantum mechanics, the linear combination $|\psi\rangle = \alpha|0\rangle + \beta|1\rangle$ with the complex numbers $\alpha, \beta \in \mathbb{C}$ is also a (superposed) state of this Hilbert space. α and β are the amplitudes of a probability function, with $|\alpha|^2 + |\beta|^2 = 1$. This superposition is called quantum bit (qubit). Intuitively, α indicates the probability that this quantum system is in state $|0\rangle$. With the rest probability β, the quantum system is in state $|1\rangle$. In a quantum bit, there are, in general, not only two possibilities 0 and 1 such as in a classical bit but infinitely many possibilities of probabilities that one of both states, $|0\rangle$ or $|1\rangle$, can be realized.

In the case of quantum bits $|\psi\rangle = \frac{1}{\sqrt{2}}(|0\rangle + |1\rangle) = \frac{1}{\sqrt{2}}|0\rangle + \frac{1}{\sqrt{2}}|1\rangle$, it follows that $|\frac{1}{\sqrt{2}}|^2 + |\frac{1}{\sqrt{2}}|^2 = \frac{1}{2} + \frac{1}{2} = 1$, and therefore, the quantum system is with 50% in state $|0\rangle$ and 50% in $|1\rangle$. But a measurement leads to the collapse of the probability amplitude. According to the postulates of quantum mechanics, one of the two possible bit states is determined by measurement with the same probability. But that means that the result of the measurement is random. After the empirical confirmation of the Einstein–Podolsky–Rosen (EPR) experiments, it is obvious that the reason is not incomplete knowledge but principle randomness.

4.3.2. Quantum bits as state vectors of the Bloch sphere

Quantum bits can intuitively be represented as state vectors of a unit sphere with radius $r = 1$ (Bloch sphere).[71] The computation of quantum bits by the application of unitary operators corresponds to the rotation of state vectors which determine the points on the surface of the unit sphere uniquely (Fig. 4.12). The corresponding states $|\psi\rangle$ can be characterized by three cartesian space coordinates x, y, and z but also by polar coordinates of the corresponding vector with the angles θ and ϕ, and $x = \cos(\phi)\sin(\theta)$, $y = \sin(\phi)\sin(\theta)$, and $z = \cos(\theta)$.

Each vector of the Bloch sphere must satisfy the normalization condition $x^2 + y^2 + z^2 = 1$. The two poles of the sphere on the z-axis

---

[71]Benenti *et al.* (2008, p. 103).

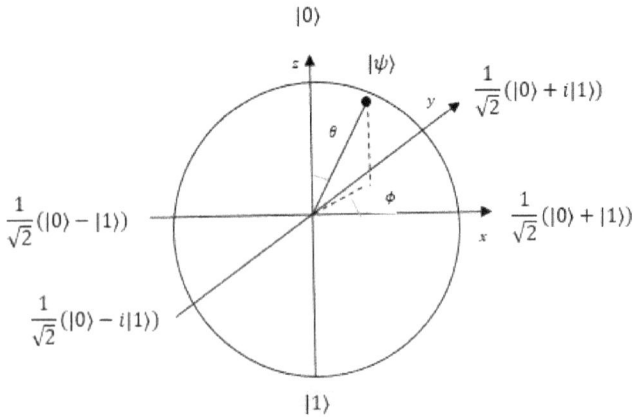

Fig. 4.12.　State vectors on the Bloch sphere.

are determined by $|0\rangle$ (north pole) and $|1\rangle$ (south pole). On the x-axis, the two superpositions $\frac{1}{\sqrt{2}}(|0\rangle + |1\rangle)$ and $\frac{1}{\sqrt{2}}(|0\rangle - |1\rangle)$ are represented. They are distinguished by a relative phase. Correspondingly, $\frac{1}{\sqrt{2}}(|0\rangle + i|1\rangle)$ and $\frac{1}{\sqrt{2}}(|0\rangle - i|1\rangle)$ can be distinguished on the y-axis.

4.3.3.　Circuits with quantum operators

Applications of unitary operators in quantum mechanics correspond to rotations of state vectors in the Bloch sphere. They are realized in a quantum computer by quantum gates, which are building blocks of the circuit in a quantum computer. Graphically, operators are represented by boxes with lines for inputs and outputs.[72]

The 1-ary classical NOT-operator corresponds to the X-operator, which flips the state $|0\rangle$ on the z-axis of the Bloch sphere into the state $|1\rangle$ and vice versa. Therefore, it is called bit-flip operator. The application of $X := \left(\begin{smallmatrix} 0 & 1 \\ 1 & 0 \end{smallmatrix}\right)$ on $|0\rangle$ delivers

$$X|0\rangle = \begin{pmatrix} 0 & 1 \\ 1 & 0 \end{pmatrix} \begin{pmatrix} 1 \\ 0 \end{pmatrix} = \begin{pmatrix} 0+0 \\ 1+0 \end{pmatrix} = \begin{pmatrix} 0 \\ 1 \end{pmatrix} = |1\rangle.$$

---

[72]Hidary (2019).

The application of the X-operator is denoted by $X|j\rangle = |j \oplus 1\rangle$ with $j = 0, 1$ and $\oplus$ as addition modulo 2. The corresponding circuit with application on $|0\rangle$ is graphically denoted in the following way.

$$|0\rangle \text{————} \oplus \text{————} |1\rangle$$

The Y-operator rotates the state vector around the y-axis. $Y := \begin{pmatrix} 0 & -i \\ i & 0 \end{pmatrix}$ applied to $|1\rangle$ delivers the state

$$Y|1\rangle = \begin{pmatrix} 0 & -i \\ i & 0 \end{pmatrix} \begin{pmatrix} 0 \\ 1 \end{pmatrix} = \begin{pmatrix} 0 - i \\ 0 + 0 \end{pmatrix} = \begin{pmatrix} -i \\ 0 \end{pmatrix} = -i|0\rangle.$$

The Y-operator is denoted as a quantum gate in the circuit of a quantum computer as follows.

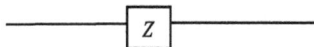

$$\text{————}\boxed{Y}\text{————}$$

The Z-operator rotates the state vector around the z-axis. It is also called phase-flip operator because it flips the state vector with radian π, resp.180°. The application of $Z := \begin{pmatrix} 1 & 0 \\ 0 & -1 \end{pmatrix}$ to $|0\rangle$ yields

$$Z|0\rangle = \begin{pmatrix} 1 & 0 \\ 0 & -1 \end{pmatrix} \begin{pmatrix} 1 \\ 0 \end{pmatrix} = \begin{pmatrix} 1 + 0 \\ 0 + 0 \end{pmatrix} = \begin{pmatrix} 1 \\ 0 \end{pmatrix} = |0\rangle.$$

The application of Z to $|1\rangle$ yields

$$Z|1\rangle = \begin{pmatrix} 1 & 0 \\ 0 & -1 \end{pmatrix} \begin{pmatrix} 0 \\ 1 \end{pmatrix} = \begin{pmatrix} 0 + 0 \\ 0 - 1 \end{pmatrix} = \begin{pmatrix} 0 \\ -1 \end{pmatrix} = -|1\rangle.$$

The Z-operator is denoted as a quantum gate in the circuit of a quantum computer as follows.

$$\text{————}\boxed{Z}\text{————}$$

Together with the identity matrix and the $\mp$- and i-multiples, they form the Pauli group.

The general phase-shift operator R_φ lets the state $|0\rangle$ unchanged, but it rotates the state $|1\rangle$ with angle resp. shift φ with $R_\varphi :=$ $\begin{pmatrix} 1 & 0 \\ 0 & e^{i\varphi} \end{pmatrix}$. The R_φ-operator is denoted as a quantum gate in the circuit of a quantum computer as follows.

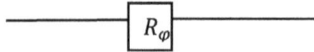

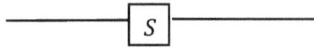

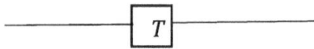

$$\boxed{R_\varphi}$$

The Pauli operator Z is a special case of R_φ for the shift $\varphi = \pi$. Because of the Euler identity $e^{i\pi} = -1$, $e^{i\pi}$ can be replaced by -1 in the Z-matrix.

Two further phase shift operators are also the special cases of the rotation operators R_φ: For $\varphi = \pi/2$, the S-Operator $S :=$ $\begin{pmatrix} 1 & 0 \\ 0 & i \end{pmatrix}$ rotates the state around the z-axis by $90°$. The S-operator is denoted as a quantum gate in the circuit of a quantum computer as follows.

$$\boxed{S}$$

For $\varphi = \pi/4$, the T-operator $T := \begin{pmatrix} 1 & 0 \\ 0 & e^{i\pi/4} \end{pmatrix}$ rotates the state around the z-axis by $45°$. Obviously, $T^2 = S$ because both the rotations by $45°$ successively yields $90° = 45° + 45°$. The T-operator is denoted as a quantum gate in the circuit of a quantum computer as follows.

$$\boxed{T}$$

An important gate is Hadamard operator H because it transforms a quantum bit into a superposition of two states. The application of $H := \frac{1}{\sqrt{2}} \begin{pmatrix} 1 & 0 \\ 0 & -1 \end{pmatrix}$ to $|0\rangle$ yields

$$H|0\rangle = \frac{1}{\sqrt{2}} \begin{pmatrix} 1 & 0 \\ 0 & -1 \end{pmatrix} \begin{pmatrix} 1 \\ 0 \end{pmatrix} = \frac{1}{\sqrt{2}} \begin{pmatrix} 1+0 \\ 1+0 \end{pmatrix}$$

$$= \frac{1}{\sqrt{2}} \begin{pmatrix} 1 \\ 1 \end{pmatrix} = \frac{1}{\sqrt{2}}(|0\rangle + |1\rangle).$$

The application of H to $|1\rangle$ yields

$$H|1\rangle = \frac{1}{\sqrt{2}} \begin{pmatrix} 1 & 0 \\ 0 & -1 \end{pmatrix} \begin{pmatrix} 0 \\ 1 \end{pmatrix} = \frac{1}{\sqrt{2}} \begin{pmatrix} 0+1 \\ 0-1 \end{pmatrix}$$

$$= \frac{1}{\sqrt{2}} \begin{pmatrix} 1 \\ -1 \end{pmatrix} = \frac{1}{\sqrt{2}}(|0\rangle - |1\rangle).$$

The H-operator is denoted as a quantum gate in the circuit of a quantum computer as follows.

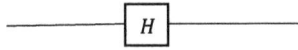

With the identity operator $I := \begin{pmatrix} 1 & 0 \\ 0 & 1 \end{pmatrix}$, the introduced 1-ary operators satisfy the equations $HXH = Z, HZH = X, HYH = -Y, H^\dagger = H$, and $H^2 = I$.

The square root of the X-, resp. NOT-gate, corresponds to a 1-ary operator which is represented by the following matrix:

$$\sqrt{X} = \frac{1}{2} \begin{pmatrix} 1+i & 1-i \\ 1-i & 1+i \end{pmatrix}.$$

It follows that

$$X = (\sqrt{X})^2 = \frac{1}{2} \begin{pmatrix} 1+i & 1-i \\ 1-i & 1+i \end{pmatrix} \frac{1}{2} \begin{pmatrix} 1+i & 1-i \\ 1-i & 1+i \end{pmatrix}$$

$$= \frac{1}{4} \begin{pmatrix} 0 & 4 \\ 4 & 0 \end{pmatrix} = \begin{pmatrix} 0 & 1 \\ 1 & 0 \end{pmatrix}.$$

This gate transforms the quantum state $|0\rangle$ into $\frac{1}{2}((1+i)|0\rangle + (1-i)|1\rangle)$ and $|1\rangle$ into $\frac{1}{2}((1-i)|0\rangle + (1+i)|1\rangle)$. As quantum gate, this operator is denoted as follows.

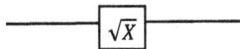

Square root gates can be generated for all gates. Therefore, a unitary matrix must be found which, multiplied with itself, corresponds to the desired gate.

In the next step, 2-ary operators for 2-ary qubits are introduced. The $SWAP$-operator swaps the 2-ary qubit $|01\rangle$ with $|10\rangle$ and vice versa. The basis states of the Hilbert space are determined by the vectors

$$|00\rangle = \begin{pmatrix} 1 \\ 0 \\ 0 \\ 0 \end{pmatrix}, \quad |01\rangle = \begin{pmatrix} 0 \\ 1 \\ 0 \\ 0 \end{pmatrix}, \quad |10\rangle = \begin{pmatrix} 0 \\ 0 \\ 1 \\ 0 \end{pmatrix}, \quad \text{and} \quad |11\rangle = \begin{pmatrix} 0 \\ 0 \\ 0 \\ 1 \end{pmatrix}.$$

After application to $|01\rangle$, the $SWAP$-operator yields

$$SWAP|01\rangle = \begin{pmatrix} 1 & 0 & 0 & 0 \\ 0 & 0 & 1 & 0 \\ 0 & 1 & 0 & 0 \\ 0 & 0 & 0 & 1 \end{pmatrix} \begin{pmatrix} 0 \\ 1 \\ 0 \\ 0 \end{pmatrix} = \begin{pmatrix} 0+0+0+0 \\ 0+0+0+0 \\ 0+1+0+0 \\ 0+0+0+0 \end{pmatrix}$$

$$= \begin{pmatrix} 0 \\ 0 \\ 1 \\ 0 \end{pmatrix} = |10\rangle.$$

The $SWAP$-operator is denoted as a quantum gate in the circuit of a quantum computer as follows.

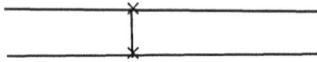

A central role in quantum computing is played by the controlled NOT $(CNOT)$–operator, which can entangle two qubits. In this 2-ary operator, the first qubit is understood as a control qubit and the second one as a target qubit. If the control qubit is in state $|0\rangle$, then the target qubit remains unchanged. If the control qubit is in state $|1\rangle$, then the NOT-operator X is applied to the target qubit. The $CNOT$-gate is used to generate entangled states. For example,

$CNOT$ applied to $|01\rangle$ yields

$$CNOT|01\rangle = \begin{pmatrix} 1 & 0 & 0 & 0 \\ 0 & 1 & 0 & 0 \\ 0 & 0 & 0 & 1 \\ 0 & 0 & 1 & 0 \end{pmatrix} \begin{pmatrix} 0 \\ 0 \\ 1 \\ 0 \end{pmatrix} = \begin{pmatrix} 0+0+0+0 \\ 0+0+0+0 \\ 0+0+0+0 \\ 0+0+1+0 \end{pmatrix}$$

$$= \begin{pmatrix} 0 \\ 0 \\ 0 \\ 1 \end{pmatrix} = |11\rangle.$$

The $CNOT$-operator is denoted as a quantum gate in the circuit of a quantum computer as follows.

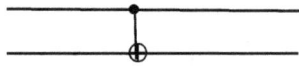

Another control operator CZ applies the Z-operator to the target qubit under the conditions of the $CNOT$-operator:

$$CZ := \begin{pmatrix} 1 & 0 & 0 & 0 \\ 0 & 1 & 0 & 0 \\ 0 & 0 & 1 & 0 \\ 0 & 0 & 0 & -1 \end{pmatrix}.$$

Contrary to the $CNOT$-operator, the control and target qubits can be exchanged, i.e., the CZ-gate is symmetric. Therefore, in the graphical representation of the circuit, the two lines are denoted with a point as follows.

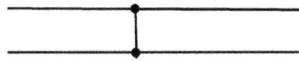

After the 1- and 2-ary operators, 3-ary operators are introduced: One example is the Toffoli operator with two control qubits and a target qubit. Therefore, it is also called $CCNOT$-operator. In order to change the target qubit, both control qubits must be in the state $|1\rangle$. The first two qubits x and y satisfy the classical (Boolean) AND-function in order to apply the NOT-function to the target qubit in the case of $x = 1$ and $y = 1$:

$$(x, y, z) \mapsto (x, y, (z \otimes xy)).$$

Applied to $|110\rangle$, the $CCNOT$-operator yields

$$CCNOT|110\rangle = \begin{pmatrix} 1 & 0 & 0 & 0 & 0 & 0 & 0 & 0 \\ 0 & 1 & 0 & 0 & 0 & 0 & 0 & 0 \\ 0 & 0 & 1 & 0 & 0 & 0 & 0 & 0 \\ 0 & 0 & 0 & 1 & 0 & 0 & 0 & 0 \\ 0 & 0 & 0 & 0 & 1 & 0 & 0 & 0 \\ 0 & 0 & 0 & 0 & 0 & 1 & 0 & 0 \\ 0 & 0 & 0 & 0 & 0 & 0 & 0 & 1 \\ 0 & 0 & 0 & 0 & 0 & 0 & 1 & 0 \end{pmatrix} \begin{pmatrix} 0 \\ 0 \\ 0 \\ 0 \\ 0 \\ 0 \\ 1 \\ 0 \end{pmatrix} = \begin{pmatrix} 0 \\ 0 \\ 0 \\ 0 \\ 0 \\ 0 \\ 0 \\ 1 \end{pmatrix} = |111\rangle.$$

In the graphical representation of the circuit, the Toffoli gate is illustrated with two lines for both control qubits and a line for the target qubit as follows.

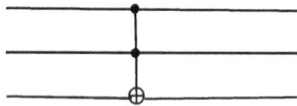

Another central 3-ary operator for quantum computing is the Fredkin operator, which is also called the controlled $SWAP$-operator, i.e., $CSWAP$-operator. In this case, the first qubit is designated as the control qubit and the other two qubits are designated as the target qubits. If the control qubit is in the state $|0\rangle$, the target qubits remain unchanged. However, if the control qubit is $|1\rangle$, then the two target qubits are exchanged in the sense of the $SWAP$-operator. In the graphical representation of the circuit model, the Fredkin gate is shown with one line for the control qubit and two lines for the two target qubits as follows.

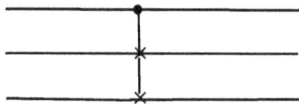

4.3.4. No cloning theorem

A classical computer can copy information and thus store it during calculations. In the circuit model of the classical computer, the $COPY$-gate was introduced for this purpose. Is there an analogous quantum gate? A quantum copier is impossible within the framework of quantum mechanics. If there were a quantum copier, then one could produce two copies of the state of a quantum system in order to measure the momentum of one copy and separately the location of the other copy with arbitrary accuracy. However, this is impossible according to the Heisenberg's uncertainty relation. The so-called no cloning theorem has far-reaching technical consequences. One of the consequences is that data backup through copies, as we know it from the classical computer, is impossible.[73]

4.3.5. Unitary transformations

Unitary transformations in quantum mechanics are reversible. This does not generally apply to the gates of a classical circuit. For example, the classical AND-gate assigns the output 0 to all three input pairs 00, 01, and 10. In the output state 0, the gate has therefore "forgotten" which input pair was actually assumed. The gate would therefore have to be equipped with a kind of "memory" in order to output the two respective input values in the output state. In the given example of the AND-gate, this would be one of the three possible output triples 0 01, 0 10, or 0 11. This would translate a non-reversible (irreversible) calculation into a reversible (reversible) calculation.

Another example is the XOR-gate. Here, an output of two output values is sufficient, from which the result of the XOR application and the two respective input values can be clearly deduced. The assignment rule of the two input and output values is $(x, y) \longmapsto (x, x \oplus y)$, with the addition $\oplus$ (modulo 2) for bits. If the first input value is 0, as in the cases 00 and 01, the output values remain unchanged: 00 and 01. If the first input value is 1, as in the cases 10 and 11, the first output value remains 1 and the second output value takes the opposite value, i.e., 11 for 01 and 10 for 11. According to this rule, it

---

[73]Homeister (2018, p. 83).

can be clearly inferred that, for example, for output 11 the input was 10, and correspondingly, for output 10 the input was 11, for output 01 the input was 01, and for output 00 the input was 00.

In fact, it can be shown, in general, that any classical calculation can be translated into a reversible calculation. Any classical invertible computation can then be simulated by a quantum circuit with unitary transformations. In this sense, any classical computer can be simulated by a quantum computer.

Translations of classical circuits in quantum circuits, however, must be realized by unitary transformations. According to Church's thesis, a function f is computable if there is a circuit that computes the function value $f(x)$ for each input x. For quantum circuits this means: Three registers $|x\rangle|h\rangle|0\rangle$ with input $|x\rangle$, the constant auxiliary bits $|h\rangle$, and an empty register $|0\rangle$ must be assigned the result $|x\rangle|h^x\rangle|f(x)\rangle$. Here, h^x indicates that auxiliary bits can depend on the input x. Since every reversible operation is a permutation of the input bits, it is also unitary. In general, a superposition $\sum_x \alpha_x|x\rangle|0\rangle$ of inputs by a quantum circuit of a quantum computer can be assigned to an entangled state $\sum_x \alpha_x|x\rangle|f(x)\rangle$ of results.

4.3.6. *Computation by circuits of a quantum computer*

Analogous to a classical circuit, a quantum circuit is composed of quantum lines and quantum gates. One quantum bit is processed in each quantum line. Quantum lines connect quantum gates that perform unitary transformations. A quantum circuit Q thus realizes a unitary transformation of an input $|\psi\rangle$ into an output $Q|\psi\rangle$ composed of quantum gates, as shown in the following.

$$|\psi\rangle \quad\boxed{\quad Q \quad}\quad Q|\psi\rangle$$

Since quantum circuits perform reversible computations as unitary transformations, the number of output channels is equal to the number of input channels. Quantum circuits consisting of parallel lines are represented by tensor products of quantum operators corresponding to quantum gates. A simple example is the following circuit

$Q = M(I_2 \otimes K)$ with the two quantum operators K and M, and the identity operator I_2. Here, Q for the input $|\psi\rangle = |x_1 x_0\rangle$ yields the output $Q|\psi\rangle = |x_1' x_0'\rangle$ as follows.

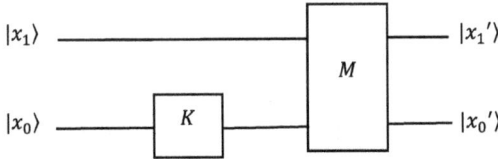

In summary, the quantum computation of a quantum computer starts with the input of a quantum register $R = |x_{n-1} \cdots x_1 x_0\rangle$, which consists of n quantum bits and is represented by a 2^n-dimensional vector space over the complex numbers. The set of vectors corresponding to the numbers $0, 1, \ldots, 2^n - 1$ and represented in binary by $|0 \ldots 00\rangle, |0 \ldots 01\rangle, \ldots, |1 \ldots 11\rangle$, with $|i\rangle$ for $i \in \{0, 1\}^n$, is chosen as the basis of this vector space.

The state of a quantum register R is a vector $\begin{pmatrix} \alpha_0 \\ \vdots \\ \alpha_{2^n-1} \end{pmatrix}$ of length 1. This state vector represents a superposition $R = \sum_{i=0}^{2^n-1} \alpha_i |i\rangle$ with the normalization condition $\sum_{i=0}^{2^n-1} |\alpha_i|^2 = 1$. The computational steps performed on quantum registers are unitary transformations and thus reversible. They can be decomposed into unitary transformations in which, at most, two bits are involved. This decomposition corresponds to tensor products.[74]

4.3.7.   *Reading out the results of a quantum computer*

After the calculation by a quantum computer comes the reading of the results. According to the postulates of quantum mechanics, reading out the results corresponds to measuring a quantum register in the state $R = \sum_{i=0}^{2^n-1} \alpha_i |i\rangle$. If this state is measured with respect to the basis $|0\rangle|, 1\rangle, \ldots, |2^n - 1\rangle$, the result $|i\rangle$ is obtained with probability $|\alpha_i|^2$. Choosing an orthonormal basis $|a_0\rangle, |a_1\rangle, \ldots, |a_{2^n-1}\rangle$ yields a measurement of the register state $R = \sum_{i=0}^{2^n-1} \alpha_i' |a_i\rangle$ with probability $|\alpha_i'|^2$ of the result $|a_i\rangle$. The measurements of observables yield

---

[74]Benenti *et al.* (2008 p. 106), Homeister (2018, p. 52).

real numbers as eigenvalues. In the case of the quantum computer, the measurement ("readout") thus yields classical information as a result.[75]

In quantum physics, a measurement is not a unitary (reversible) transformation that transforms a superposition into another superposition. According to the postulates of quantum mechanics, a superposition rather "decays" into a probability distribution of its partial states. In quantum mechanics, this subsequent state is also called a "mixed" state in contrast to the "pure" state of a superposition. In a quantum circuit, a special symbol is used for this purpose as the termination of a calculation. The following diagram is considered as an example of a circuit with measurement termination.

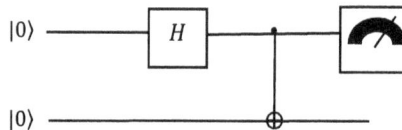

As an example of a quantum circuit with measurement termination, two qubits ψ_1 and ψ_2 in the state $|0\rangle$ are to be prepared first. An application of the Hadamard operator H to the first qubit transforms it into a superposition of the states $|\psi_1\rangle = \frac{1}{\sqrt{2}}(|0\rangle + |1\rangle)$. Then, the $CNOT$-operator is applied to ψ_1 and ψ_2. This transforms the two qubits into an entangled state $\frac{1}{\sqrt{2}}(|00\rangle + |11\rangle)$, which is not separable. In the sense of quantum mechanics, this is an EPR (Bellian) state. Then, ψ_1 is measured with a 50% chance of obtaining 0 or 1 as a real output (eigenvalue).

4.3.8.   *Architecture of quantum circuits*

To establish the proximity of a quantum circuit to the technology of a quantum computer, a three-dimensional model can be imagined, in which the processing of quantum information proceeds in layers of quantum gates from left to right. This illustrates the depth of the quantum circuit. In the first input layer, the qubits in the prepared state (indicated by up and down arrows for alternative bit states) are shown on the left. In the second layer, Hadamard gates are used

---

[75]Hidary (2019 p. 34), Homeister (2018, pp. 29, 46).

to transform the input into a superposition. In the following layers, 1- and 2-qubit gates, such as X, Y, T, and CZ are used. A measurement then completes the calculation in the quantum circuit.

After measurement in a quantum circuit, classical bits can be read out as classical information, which can be used for further processing by classical computers, for example. According to the postulates of quantum mechanics, the measurement is based on the probability amplitudes of the quantum states of each qubit. A central question of quantum computing is, therefore, how to predetermine which of the qubits is the result of the calculation. Quantum algorithms, such as the quantum Fourier transform, have been proposed for this purpose.

Superposition and entanglement open up completely new possibilities for quantum algorithms that are closed to classical algorithms. A classical computer with its classical gates can only solve one task at a time. If several tasks are to be solved, then this can only happen one after the other and not simultaneously (in parallel). In classical supercomputers, one also speaks of "parallelism." In this case, different tasks are solved very quickly one after the other by a circuit network, or separate circuit networks are used that solve different subtasks simultaneously. For technical reasons, however, the possibilities for building separate parallel classical circuit networks are limited.

In a quantum computer, in principle, any number of input data from subtasks can be superimposed in a superposition and then transferred by a single application of a quantum switching network into a superposition that contains all possible solutions. Figure 4.13 illustrates for the example of an input sequence of four bits how all $2^4 = 16$ possibilities are combined in a superposition. In general, a sequence of n bits leads to a superposition with 2^n possibilities. However, reading out these possible solutions involves a quantum-mechanical measurement in which the superposition collapses and one of the possible solutions is randomly output with equal probability. If one wanted to try out all possible results in the superposition through brute force with classical algorithms, the computing time would "shoot through the roof" exponentially depending on the size of the superposition. How can the probability of the solution sought be influenced before it is triggered? How can errors be minimized in the process? These are the challenges of effective selection of algorithms.

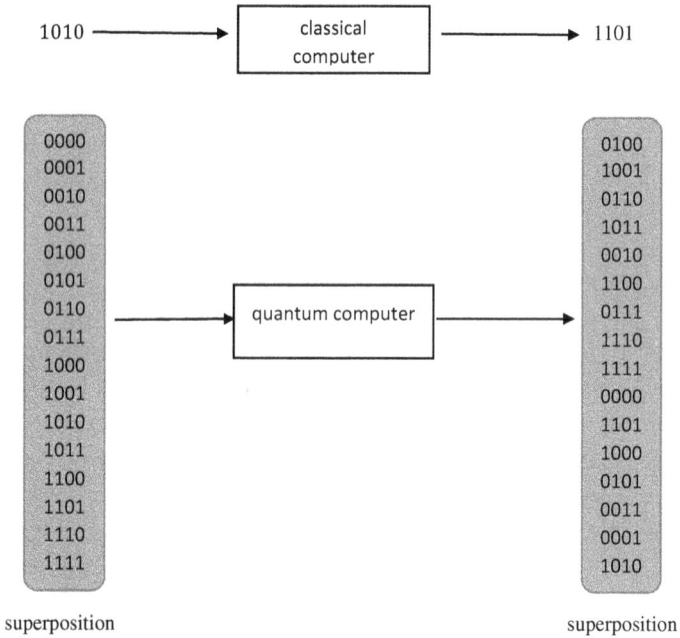

Fig. 4.13.   Superposition and quantum parallelism.

4.3.9.   *What does entanglement mean?*[76]

Besides superposition, another important feature of the quantum world, which makes modern quantum technology possible, is the entanglement of quantum states.

In composite quantum systems, entangled states are observed, which, along with superpositions, are among the fundamental differences between quantum physics, classical physics, and our everyday understanding. In addition, entangled systems, along with superpositions, are fundamental to quantum computing. Behind this is the observation that quantum systems can remain correlated even if they are spatially separated and without physical interaction. Einstein called this kind of "action in distance" "spooky" from the standpoint of classical physics. In fact, however, the quantum-mechanical formalism allows for these phenomena.

---

[76]Mainzer (2020, pp. 47–50).

Several quantum systems can be combined into one overall system. If two quantum systems are described by the state vectors of two Hilbert spaces $\mathcal{H}_1$ and $\mathcal{H}_2$, then the entire system is represented by the tensor product $\mathcal{H} = \mathcal{H}_1 \otimes \mathcal{H}_2$. According to the definition of the tensor product, most states in the Hilbert space $\mathcal{H}$ do not consist of tensor products of states from $\mathcal{H}_1$ and $\mathcal{H}_2$ but of their superpositions. States from $\mathcal{H} = \mathcal{H}_1 \otimes \mathcal{H}_2$ are written as $|\psi\rangle = \sum_{ij} c_{ij} |ij\rangle$, where index i refers to a state from $\mathcal{H}_1$ and j to a state from $\mathcal{H}_2$.

A state $|\psi\rangle$ in the Hilbert space $\mathcal{H} = \mathcal{H}_1 \otimes \mathcal{H}_2$ of a quantum system is called entangled or non-separable if $|\psi\rangle$ cannot be represented as a tensor product with a state $|\alpha\rangle_1$ from $\mathcal{H}_1$ and a state $|\alpha\rangle_2$ from $\mathcal{H}_2$, i.e., $|\psi\rangle \neq |\alpha\rangle_1 \otimes |\alpha\rangle_2$.

A state $|\psi\rangle$ in the Hilbert space $\mathcal{H} = \mathcal{H}_1 \otimes \mathcal{H}_2$ of a quantum system is called separable if there is a state $|\alpha\rangle_1$ of $\mathcal{H}_1$ and a state $|\alpha\rangle_2$ of H_2 with $|\psi\rangle = |\alpha\rangle_1 \otimes |\alpha\rangle_2$.

An example of entanglement is the state $|\psi_1\rangle = \frac{1}{\sqrt{2}}(|00\rangle + |11\rangle)$. In contrast, the state $|\psi_2\rangle = \frac{1}{\sqrt{2}}(|01\rangle + |11\rangle)$ is separable because $|\psi_2\rangle = \frac{1}{\sqrt{2}}(|0\rangle + |1\rangle) \otimes |1\rangle$. If two systems are entangled, then they cannot be distinguished with individual state vectors of the two subsystems. In everyday life and in classical physics, however, we naturally assume that all bodies, such as the planets, belong to an overall system, such as the planetary system, but can be described individually with separate states and properties. What is the physical significance of these consequences of the formalism of quantum mechanics?

For this purpose, we consider polarization experiments with photons. A polarization filter lets linearly polarized light through in the horizontal direction but not in the vertical direction. The transmission is indeterminate for polarization angles between $0°$ and $90°$. A photon source can send photon pairs in opposite directions to polarization filters P_1 and P_2. Mathematically, the states of photons 1 and 2 are represented by the Hilbert spaces $\mathcal{H}_1$ and $\mathcal{H}_2$, respectively.

Let $|\psi_1\rangle$ and $|\psi_2\rangle$ be states from $\mathcal{H}_1$ with an observable A and eigenvalues (measured values) a_1 and a_2, respectively. Let $|\varphi_1\rangle$ and $|\varphi_2\rangle$ be states from $\mathcal{H}_2$ with an observable B and eigenvalues (measured values) b_1 and b_2s respectively. Then, in the composite system

$$\mathcal{H} = \mathcal{H}_1 \otimes \mathcal{H}_2,$$

$$|\psi\rangle = \frac{1}{\sqrt{2}}(|\psi_1\rangle \otimes |\varphi_1\rangle + |\psi_2\rangle \otimes |\varphi_2\rangle)$$

is also a state which, however, is not separable. In the entangled $|\psi\rangle$ state, the subsystems are correlated, i.e., in the states $|\psi_1\rangle \otimes |\varphi_1\rangle$, respectively $|\psi_2\rangle \otimes |\varphi_2\rangle$, the observables A and B of the subsystems can have the values a_1 and b_1 or a_2 and b_2 with equal probability of 1:2 but never a_1 and b_2 or a_2 and b_1. The photon pairs $|\psi_1\rangle \otimes |\varphi_1\rangle$, resp. $|\psi_2\rangle \otimes |\varphi_2\rangle$, indicate the states of the photon pair in the vertical or horizontal direction linearly polarized. The overall state $|\psi\rangle$ is now interpreted as the correlation of the polarization of the two photons.

Since $|\psi\rangle$ contains the states $|\psi_1\rangle \otimes |\varphi_1\rangle$ and $|\psi_2\rangle \otimes |\varphi_2\rangle$ in equal parts, the probability that both photons are transmitted or not transmitted is 50% in each case. It is excluded that one photon is transmitted and the other is not. One can then say that both photons are strictly correlated, i.e., they behave in strict agreement, although they are spatially separated and without physical interaction.

Quantum mechanics thus enables correlated ("entangled") total states of systems whose partial states cannot be "localized." How can this be reconciled with the "local" realism of classical physics, according to which physical systems possess, at least in principle, well-determined properties at every point in time and in every state, independent of observation and measurement? In 1964, assuming Einstein's reality and locality principle, John Bell used classical statistics to derive an inequality that takes into account all possible states in the EPR experiment.[77] From this, predictions can be derived that significantly contradict empirical measurement results in the quantum-mechanical execution of the EPR experiment.[78]

4.3.10.   *Entanglement and quantum communication*

Entanglement enables quantum communication. A teleportation can transmit a quantum bit using a classical information channel if the

---

[77]Bell (1964).
[78]Aspect *et al.* (1981).

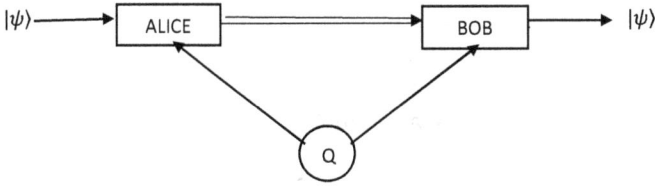

Fig. 4.14. Quantum teleportation with EPR source Q.

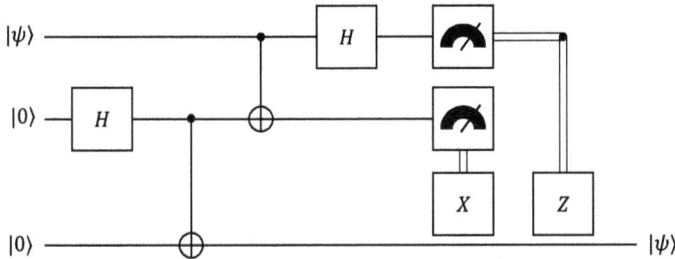

Fig. 4.15. Quantum circuit of quantum teleportation.[79]

communication partners share an entangled bit pair. In Fig. 4.14, Alice sends a qubit in the quantum state $|\psi\rangle$ to Bob, but she has only a classical channel with two classical bits (shown by a double line) available for this purpose, and she shares an entangled state with Bob via an EPR source Q. She entangles the quantum bit to be transmitted with her part of the EPR pair. Then, she measures her bits and instantaneously changes Bob's quantum bit. This is not a contradiction to relativity because Alice's and Bob's parts are entangled in a common quantum state. Finally, Alice sends her measurement result to Bob using the classical channel.[80]

For the quantum teleportation in Fig. 4.14, a corresponding quantum circuit can be constructed, as shown in Fig. 4.15. Alice has a quantum bit $|x\rangle$ in the state $|\psi\rangle = \alpha|0\rangle + \beta|1\rangle$, which she wants to send to Bob. In addition, she has a quantum bit $|a\rangle$, while Bob has a quantum bit $|b\rangle$. Both quantum bits are to be prepared in an entangled state $|ab\rangle = \frac{1}{\sqrt{2}}(|00\rangle + |11\rangle)$. As in Fig. 4.14, an EPR source is

---

[79]Benenti *et al.* (2008, p. 209).
[80]Keyl (2002, Chapter 4.1.2).

used for this purpose, which generates an entangled photon pair, for example, and sends one photon to Alice and another to Bob.[81]

This entangled state is the Bell state $|\phi^+\rangle$, which arises from the state $|00\rangle$ by applying a Hadamard gate H and a $CNOT$-gate:

$$CNOT(H \otimes I)|00\rangle = \frac{1}{\sqrt{2}}(|00\rangle + 11\rangle) = |\phi^+\rangle.$$

Thus, the input register $|x\rangle|a\rangle|b\rangle$ with the input states $|\psi\rangle|0\rangle|0\rangle$ is in state

$$|\psi\rangle \otimes |\phi^+\rangle = (\alpha|0\rangle + \beta|1\rangle) \otimes \frac{1}{\sqrt{2}}(|00\rangle + |11\rangle)$$

$$= \frac{\alpha}{\sqrt{2}}(|000\rangle + |011\rangle) + \frac{\beta}{\sqrt{2}}(|100\rangle + |111\rangle.$$

In the next step, the $CNOT$-gate is applied, which yields

$$CNOT(|\psi\rangle \otimes |\phi^+\rangle) = \frac{\alpha}{\sqrt{2}}(|000\rangle + |011\rangle) + \frac{\beta}{\sqrt{2}}(|110\rangle + |101\rangle).$$

Then, the Hadamard gate leads to the quantum state

$$H\left(\frac{\alpha}{\sqrt{2}}(|000\rangle + 011\rangle) + \frac{\beta}{\sqrt{2}}(|110\rangle + |101\rangle)\right)$$

$$= \frac{\alpha}{\sqrt{2}}\left(\frac{1}{\sqrt{2}}(|0\rangle + |1\rangle)\right)((|00\rangle + |11\rangle)$$

$$+ \frac{\beta}{\sqrt{2}}\left(\frac{1}{\sqrt{2}}(|0\rangle - |1\rangle)\right)((|10\rangle + |01\rangle)$$

$$= \frac{\alpha}{2}(|000\rangle + |011\rangle + |100\rangle + |111\rangle)$$

$$+ \frac{\beta}{2}(|010\rangle + |001\rangle - |110\rangle - 101\rangle)$$

$$= \frac{1}{2}(|00\rangle(\alpha|0\rangle + \beta|1\rangle) + |01\rangle(\beta|0\rangle + \alpha|1\rangle)$$

$$+ |10\rangle(\alpha|0\rangle - \beta|1\rangle) - |11\rangle(\beta|0\rangle - \alpha|1\rangle)).$$

In the last transformation step, the possible result of the measurement of the first two bits can already be seen. With a probability of

---

[81]Homeister (2018, p. 129), Benenti *et al.* (2008, pp. 208–210), Mainzer (2020, p. 111).

0.25%, Alice's measurement yields one of the four possible results: $|00\rangle$, $|01\rangle$, $|10\rangle$, or $|11\rangle$. Because of the entangled state, Bob's bit instantly switches to a state correlated with Alice's result. From the equation, we can see that the $|00\rangle$ result of Alice's bit leads to the $\alpha|0\rangle + \beta|1\rangle$ state of Bob's bit, $|01\rangle$ to the state $\beta|0\rangle + \alpha|1\rangle$, $|10\rangle$ to the state $\alpha|0\rangle - \beta|1\rangle$, and $|11\rangle$ to the state $\beta|0\rangle - \alpha|1\rangle$. So, for the case $|00\rangle$ of the measurement result, Bob has obtained the quantum bit $|\psi\rangle = \alpha|0\rangle + \beta|1\rangle$.

In the other cases, $|\psi\rangle$ can be set up if the corresponding measurement results are known. These measurement results are communicated to Bob as classical information consisting of two bits via the classical channel between Alice and Bob (e.g., through telephone). In the quantum circuit of Fig. 4.15, the transmission of the two classical bits is represented by a double line. Thus, in the case of 01, Bob must exchange the amplitudes α and β of $|0\rangle$ and $|1\rangle$. This can be achieved with the quantum gate X of a bit flip. In case 10, the minus sign before the amplitude of $|1\rangle$ is achieved with the quantum gate Z of the phase flip or the Pauli matrix σ_z. In case 11, the quantum gates X and Z are to be applied.

It is clear that in the case of quantum teleportation, the information present at the receiver end at the moment (via the EPR entanglement) can only be accessed if the measurement result was previously communicated via the classical channel. However, the speed transmission of classical information is finite and at most as fast as the speed of light. Nevertheless, the instantaneous transmission of quantum bits opens up revolutionary technologies. Quantum communication can be realized by the transmission of quantum bits over long distances, e.g., using satellite technology, and over minimal distances in a quantum computer.

4.3.11. *Temporal logic for quantum state transformation*

Temporal logics are appropriate tools to describe state transformations in possible worlds of the Kripke semantics or underlying transition systems. The following temporal logic considers the transformation of single qubits in quantum worlds by means of unary quantum operators. For each qubit, unary quantum operators are used to unfold a branching-temporal model of transformations. Let

$|\psi\rangle = \alpha|0\rangle + \beta|1\rangle$ be a given qubit of a quantum system. An atomic proposition p is an encoding $\lceil|\psi\rangle\rceil$ of the mathematical description of the qubit. Concerning temporal operators, for example, the formula $E\,\Box\,p$ means that p is true at every time instant in some (E) possible future. With respect to tree-like and branching-temporal models, formulae are equipped with labels that represent quantum systems, states, quantum information, and paths. For example, $(i, x, q) : p$ means that p holds for quantum system i in state x with certain quantum information q concerning the quantum system. Intuitively, quantum bits (qubits) can be considered as agents which communicate with one another. In the following quantum branching distributed temporal logic (QBDTL), qubits are understood as agents with local states.[82] Thus, QBDTL is not an example of quantum logic[83] but can be considered a logical axiomatization of quantum theory.

The local language $\mathcal{L}_i$ of an agent i is inductively defined by

$$\mathcal{L}_i := p|\bot|\mathcal{L}_i \to \mathcal{L}_i|\mathrm{EX}\,\mathcal{L}_i|\mathrm{E}\,\Box\,\mathcal{L}_i|\mathrm{A}\,\Box\,\mathcal{L}_i|@_j\mathcal{L}_j,$$

with $p \in \mathrm{Prop}$ (class of atomic propositions), $j \in \mathrm{Id}$ (class of agent identifiers), and $j \neq i$, $\bot$ falsum, and $\to$ implication. The formula $@_j A$ (calling) means that agent i has just communicated (i.e., synchronized) with agent j for whom A holds. Formula A is built from temporal operators, synchronization, and propositional symbols with Boolean connectives, which can be defined as usual.

Following Peircean branching-temporal logic, there are only temporal operators as combinations of one single linear-time operator immediately preceded by one single path quantifier:

(i) $\mathrm{EX}A$ ("Formula A is true at the next time instant in some possible future"),

(ii) $\mathrm{E}\Box A$ ("Formula A is true at every time instant in some possible future"),

(iii) $\mathrm{A}\Box A$ ("Formula A is true at every time instant in every possible future").

The syntax of the global language QBDTL is inductively defined by

$$\mathcal{L} := @_{i_1}\mathcal{L}_{i_1}|\cdots|@_{i_n}\mathcal{L}_{i_n}|,$$

---

[82]Ehrich and Caleiro (2000), Viganò *et al.* (2015, p. 11).
[83]Dalla Chiara (1986, pp. 427–469).

with $i_1, \ldots, i_1 \in \mathrm{Id}$. The global $@_{i_k} A$ means that A is true for agent $i_k \in \mathrm{Id}$.

In agent technology, it is usual to consider the branching life cycle of an agent for temporal logic. It makes also sense to consider the "life cycle" of a qubit in the branching-temporal logic QBDTL. The branching local life cycle of an agent $i \in \mathrm{Id}$ is an ω-tree with $\lambda_i = (\mathrm{Ev}_i <_i)$, the set Ev_i of local events of i, and binary relation $<_i \subseteq \mathrm{Ev}_i \times \mathrm{Ev}_i$, which satisfies

(i) $<_i$ is transitive and irreflexive;
(ii) for each event $e \in \mathrm{Ev}_i$, the set $\{e' \in \mathrm{Ev}_i | e' <_i e\}$ is linearly ordered by $<_i$;
(iii) there is a smallest element 0_i as root of λ_i;
(iv) each maximally linearly $<_i$-ordered subset of Ev_i is order-isomorphic to the natural numbers.

The semantics of QBDTL uses tree-like event structures for agents:[84] In the branching tree of temporal logic, an immediate local successor e' of e is denoted by $e' \to_i e$ if $e' <_i e$, and there is no e'' such that $e' <_i e'' <_i e$. A sequence $(e_0, \ldots, e_n)$ of local events with $e_k \to_i e_{k+1}$ for $0 \leq k \leq n - 1$ is called a $\to_i$-path. An e-branch $b = (e_0 e_1, \ldots)$ of i is an infinite $\to_i$-path with $e = e_0$. A restriction of $\to_i$ to b is denoted by $\to_i^b$. $\mathcal{B}_i$ is the set of all $\to_i^b$. The reflexive and transitive closure of $\to_i^b$ is denoted by $\to_i^{b*}$.

A finite set $\xi \in Ev_i$ down closed for local causality is called a local state if $e <_i e'$ and $e' \in \xi$ then $e \in \xi$. Each nonempty local state ξ is reached by the occurrence of an event $last(\xi)$ from the local state $\xi \setminus \{last(\xi)\}$. For an event $e \in Ev_i$, the set $e \downarrow i = \{e' \in \mathrm{Ev}_i | e' \leq_i e\}$ with reflexive closure $\leq_i$ of $<_i$ is always a local state. For $\xi \neq \emptyset$, it is $last(\xi) \downarrow i = \xi$.

A branching distributed life cycle is a family of local life cycles $\lambda = (\lambda_i = (\mathrm{Ev}_i, <_i))_{i \in \mathrm{Id}}$ such that

(i) $\leq = (\bigcup_{i \in \mathrm{Id}} \leq_i)^*$ defines a partial order of global causality on the set $\mathrm{Ev} = \bigcup_{i \in \mathrm{Id}} \mathrm{Ev}_i$,
(ii) for $e, e' \in \mathrm{Ev}_i$: if $e \leq e'$, then $e \leq_i e'$.

---

[84]Winkel and Nielsen (1995).

The usual semantics of temporal logic with Kripke frames can be specified for qubit states.[85] An S5 Kripke frame is a structure $(\mathcal{Q}, \mathcal{U})$ with a nonempty $\mathcal{Q}$ of qubit states and a binary equivalence relation $\mathcal{U}$ on $\mathcal{Q}$ which is reflexive, symmetric, and transitive.

An S5 Kripke model is a triple $\mathfrak{M} = (\mathcal{Q}, \mathcal{U}, \mathcal{V})$ with S5 Kripke frame $(\mathcal{Q}, \mathcal{U})$ and valuation function $\mathcal{V} : \mathcal{Q} \to 2^{\text{Prop}}$ assigning to each qubit state in $\mathcal{Q}$ a set of atomic propositions.

A QBDTL model is a triple $\mu = (\lambda, \mathfrak{M}, \pi)$ with a distributed life cycle $\lambda = (\lambda_i)_{i \in \text{Id}}$, an S5 Kripke model $\mathfrak{M} = (\mathcal{Q}, \mathcal{U}, \mathcal{V})$, and a family $\pi = (\pi_i)_{i \in \text{Id}}$ of local functions which associate to each local state a qubit state in $\mathcal{Q}$. For each $i \in \text{Id}$ and set Ξ_i of local states of i, the function $\pi_i : \Xi_i \to \mathcal{Q}$ is defined as follows:

(i) If $\xi, \xi' \in \Xi_i$, $last(\xi) \to_i last(\xi')$, $\pi(\xi) = q$ and $\pi(\xi') = q'$, then $q \mathcal{U} q'$.

(ii) If $q, q' \in \mathcal{Q}$, $q \mathcal{U} q'$, and $\pi(\xi) = q$, then there exists $\xi' \in \Xi_i$ with $last(\xi) \to_i last(\xi')$ and $\pi(\xi') = q'$.

$(\lambda_i, \mathfrak{M}, \pi_i)$ is denoted by μ_i.

The semantics of the global language QBDTL is defined by the global satisfaction relation

$$\models^\mu @_i A \quad \text{iff} \quad \models_i^{\mu_i} A \quad \text{iff} \quad \models_i^{\mu_i, \xi} A \quad \text{for every } \xi \in \Xi_i,$$

with the local satisfaction relation at a local state ξ of i:

$\not\models_i^{\mu_i, \xi} \bot$;

$\models_i^{\mu_i, \xi} p$ iff $p \in \mathcal{V}(\pi_i(\xi))$ for $p \in \text{Prop}$;

$\models_i^{\mu_i, \xi} A \to B$ iff $\models_i^{\mu_i, \xi} A$ implies $\models_i^{\mu_i, \xi} B$;

$\models_i^{\mu_i, \xi} A \square A$ iff for all ξ', $last(\xi) \leq_i last(\xi')$ implies $\models_i^{\mu_i, \xi'} A$;

$\models_i^{\mu_i, \xi} E \square A$ iff there exists a $last(\xi)$-branch b such that for all ξ',

$$last(\xi) \to_i^{b^*} last(\xi') \text{ implies } \models_i^{\mu_i, \xi'} A;$$

---

[85]Viganò *et al.* (2015, pp. 12–13).

$\models_i^{\mu_i,\xi}$ EXA iff there exists ξ' such that $last(\xi) \to_i last(\xi')$ and $\models_i^{\mu_i,\xi'} A$;

$\models_i^{\mu_i,\xi}$ @$_j A$ iff $last(\xi) \in \text{Ev}_i$ and $\models_i^{\mu_j,last(\xi)\uparrow j} A$.

The following abbreviations are purposeful for sets formulae:

$$\models^\mu \Gamma \quad \text{iff} \quad \models^\mu A \quad \text{for all } A \in \Gamma,$$

$$\Gamma \models^\mu A \quad \text{iff} \quad \models^\mu \Gamma \quad \text{implies} \quad \models^\mu A,$$

$$\Gamma \models A \quad \text{iff} \quad \Gamma \models^\mu A \quad \text{for each QBDTL model } \mu.$$

After the definition of syntax and semantics of the temporal logic QBDTL, the set Prop of proposition symbols can be introduced to describe the behavior of quantum gates. Each propositional symbol $p \in$ Prop is the encoding $\lceil |\varphi\rangle \rceil$ of a qubit $|\varphi\rangle \in Q$ in the QBDTL syntax. The set Prop of propositional symbols contains the encodings $\lceil |\alpha|0\rangle + \beta|1\rangle \rceil$ as well as $\lceil U_1(U_2(\cdots(U_n(|\alpha|0\rangle + \beta|1\rangle)))\cdots)) \rceil$, with U_j either a unitary transformation or the restriction of a *CNOT*-operator. Different propositional symbols can describe equivalent quantum states, such as $\lceil H|0\rangle \rceil$ and $\lceil \frac{1}{2}(||0\rangle+|1\rangle) \rceil$ with the Hadamard gate H and the same quantum state.

The simplest quantum circuit that acts on two agents is built upon a single occurrence of the *CNOT*-gate. In Fig. 4.16, C is an instance of the *CNOT*-gate with control input i and target input j, which are set to $|1\rangle$ and $|0\rangle$, respectively. The *CNOT*-gate realizes the negation of the target qubit. The time instants t_1 and t_2 refer to

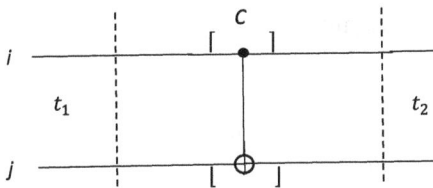

Fig. 4.16.   Instance C of the *CNOT*-gate with control input i and target input j.[86]

---

[86]Viganò *et al.* (2015, p. 14).

the temporal evolution of the inputs $|1\rangle$ and $|0\rangle$[87]:

$$t_1 : \quad @_i(\lceil|1\rangle\rceil) \text{ and } @_j(\lceil|0\rangle\rceil), \tag{1}$$

$$t_1 : \quad @_i(\lceil|1\rangle\rceil \wedge \text{EX}\lceil|1\rangle\rceil), \tag{2}$$

$$t_1 : \quad @_j(\lceil|0\rangle\rceil \wedge @_i(\lceil|1\rangle\rceil)) \rightarrow \text{EX}\lceil|1\rangle\rceil), \tag{3}$$

$$t_2 : \quad @_i(\lceil|1\rangle\rceil) \text{ and } @_j(\lceil|1\rangle\rceil). \tag{4}$$

Formula (1) is the input at time instant t_1. Formula (2) means that the control qubit of the $CNOT$-gate, resp. the first agent, has input $|0\rangle$, and there exists a successor temporal state in which the state remains unchanged because the $CNOT$-operator does not act on the control qubit. Formula (3) describes the occurrence of the binary gate which corresponds to the control, resp. synchronization, between agents. Thus, the formula means that a control or QBDTL calling, and the related operation occur as a consequence of agent synchronization. Formula (4) is the output.

In the branching-temporal logic QBDTL, derivations should be realized in a Gentzen-style with natural deductions. In order to formalize the system $\mathcal{N}(\text{QBDTL})$ of natural deductions, the formulae are equipped with labels referring to agents, states, quantum information, and paths.[88] In $(i, x, q) : A$, the label (i, x, q) refers to agents $i \in \text{Id}$, local states $x \in \text{Lab}_S$ of agents, and quantum information $q \in \text{Lab}_Q$ of agents. Further labels $\lhd, \lhd_1, \lhd_2, \ldots \in \text{Lab}_B$ refer to the successor relation between the local states in the local life cycle of an agent i along a given branch.

The semantics of QBDTL must be extended by an interpretation of the labels. Without going into the details, for a given QBDTL model μ, an interpretation function of labels is a triple $\mathcal{J} = (\mathcal{J}_S, \mathcal{J}_Q, \mathcal{J}_B)$, with, for example, a set $\mathcal{J}_S = (\mathcal{J}_S^i)_{i \in \text{Id}}$ of functions $\mathcal{J}_S^i : \text{Lab}_S \rightarrow \Xi_i$ for each $i \in \text{Id}$ and $\mathcal{J}_Q : \text{Lab}_Q \rightarrow Q$. For a QBDTL model μ and an interpretation function I, the truth of a labeled formula is defined by

$$\models^{\mu,\mathcal{J}} (i, x, q) : A \quad \text{iff} \quad \mu_i, \mathcal{J}_S^i(x) \models_i A \quad \text{and} \quad \pi_i(\mathcal{J}_S^i(x)) = \mathcal{J}_Q(q).$$

---

[87]Viganò *et al.* (2015, pp. 14–15).
[88]Gabbay (1996), Viganò (2000), Basin *et al.* (2011).

Again, there are abbreviations, such as

$\models^{\mu,\mathcal{J}} \Gamma$ iff $\models^{\mu,\mathcal{J}} \gamma$ for all $\gamma \in \Gamma$,

$\Gamma \models^{\mu,\mathcal{J}} \gamma$ iff $\models^{\mu,\mathcal{J}} \Gamma$ implies $\models^{\mu,\mathcal{J}} \gamma$,

$\models^{\mu} \gamma$ iff $\models^{\mu,\mathcal{J}} \gamma$ for every interpretation function $\mathcal{J}$,

$\models^{\mu} \Gamma$ iff $\models^{\mu,\mathcal{J}} \Gamma$ for every interpretation function $\mathcal{J}$,

$\Gamma \models \gamma$ iff $\Gamma \models^{\mu,\mathcal{J}} \gamma$ for every QBDTL $-$ model $\mathfrak{M}$

            and interpretation function $\mathcal{J}$.

The rules of natural deductions in $\mathcal{N}$(QBDTL) are given in the Gentzen style. They can be classified as local life-cycle rules, distributed life-cycle rules, quantum transformations rules, and interaction rules. The local life-cycle rules infer formulae which are local to an agent i with label s_i. These rules refer to classical connectives with elimination (E) and introduction (I) rules, $\bot E$, $\to I$, and $\to E$, and the usual assumptions in natural deductions with brackets[89]:

$$[s_i : A \to \bot] \quad [s_i : A]$$

$$\vdots \qquad\qquad \vdots$$

$$\frac{s_j : \bot}{s_i : \bot} \bot E \qquad \frac{s_i : B}{s_i : A \to B} \to I \qquad \frac{s_i : A \to B \quad s_i : A}{s_i : B} \to E.$$

The rules for introduction and elimination of temporal operators $A\Box$, $E\Box$, and EX are similar to the corresponding rules in labeled systems of modal logic. For example, the elimination rule $A\Box E$ of $A\Box$ means that if $A\Box A$ is true in a state s_i' and s_i is accessible from s_i' along some path $(s_i' R^* s_i)$ with (closure) relation R^*, then it is possible to conclude that A is true in state s_i:

$$\frac{s_i' : A\Box A \quad s_i' R^* s_i}{s_i : A} A\Box E.$$

There are also rules for modeling properties of accessibility relations and the induction principle underlying these relations.

For distributed life cycles of agents, the rules for the calling operator @ are intuitively obvious. For example, the introduction rule

---

[89]Viganò *et al.* (2015, p. 19, Fig. 1).

$@I$ means that if agent i in state s_i synchronizes ($\bowtie$) with agent j in state s_j, and A is true for j in that state, then i just communicated with agent j[90]:

$$\frac{s_j : A \quad s_i \bowtie s_j}{s_i : @_j A} @I.$$

The quantum transformation rules model essential properties of quantum operators. The reflexivity rule $refl_U$ of unitary operator U means that the class of unitary operators includes the identity transformation. The symmetry rule $symm_U$ requires the reversibility of the quantum transformation. The transitivity rule $trans_U$ allows the composition of several unitary operators as one unitary operator. The rule *prop* means that if two composed labels (i, x, q) and (j, y, q) have the same quantum information q, then each atomic proposition that is true in (i, x, q) is also true in (j, y, q)[91]:

$$[qUq]$$
$$\vdots$$
$$\frac{s_i : A}{s_i : A} \; refl_U$$

$$[qUq'']$$
$$\vdots$$
$$\frac{qUq' \quad q'Uq'' \quad s_i : A}{s_i : A} \; trans_U$$

$$[q'Uq]$$
$$\vdots$$
$$\frac{qUq' \quad s_i : A}{s_i : A} \; symm_U$$

$$\frac{(i,x,q):p \quad \gamma(j,y,q)}{(j,y,q):p} \; prop.$$

There are also rules of natural deduction to model the interaction between U and relation R. For example, the rule $U \Rightarrow R$ means that if qUq' and the label (i, x, q) occur in the formula $\gamma(j, y, q)$, then (i, x, q) has a $\lhd$-successor (i, y, q'). The rule $R \Rightarrow U$ considers the situation that if (i, y, q') is a $\lhd^*$-successor of (i, x, q), then also the

---

[90]Viganò *et al.* (2015, p. 20, Fig. 2).
[91]Viganò *et al.* (2015, p. 20, Fig. 3).

quantum labels q and q' are related by U[92]:

$$[(i,x,q) \lhd (i,y,q')] \qquad\qquad\qquad [qUq']$$
$$\vdots \qquad\qquad\qquad\qquad\qquad \vdots$$

$$\frac{qUq' \quad \gamma(i,x,q) \qquad s_j:A}{s_j:A} \, U \Rightarrow R \qquad\qquad \frac{(i,x,q) \lhd^* (i,y,q') \quad s_j:A}{s_j:A} \, R \Rightarrow U.$$

The notions of derivation and of open and discharged assumption are defined, such as in natural deductions, as usual. The notion

$$\Gamma \vdash_{\mathcal{N}(\text{QBDTL})} (i,x,q) : A$$

means that there is a derivation of $(i,x,q) : A$ in $\mathcal{N}(\text{QBDTL})$ with open assumptions in the set Γ of formulae. A derivation of $(i,x,q) : A$ in $\mathcal{N}(\text{QBDTL})$ with all assumptions discharged is a proof of the theorem $(i,x,q) : A$.

The formal system $\mathcal{N}(\text{QBDTL})$ of natural deduction is sound wth respect to the introduced semantics, i.e., $\Gamma \vdash_{\mathcal{N}(\text{QBDTL})} (i,x,q) : A$ implies $\Gamma \models (i,x,q) : A$ for every set Γ and labeled formula $(i,x,q) : A$. The proof is straghtforward with standard tools of proof theory by induction on the structure of the derivations of $(i,x,q) : A$.

The proof of (weak) completeness of $\mathcal{N}(\text{QBDTL})$ is more ambitious, i.e., $\models (i,x,q) : A$ implies $\vdash_{\mathcal{N}(\text{QBDTL})} (i,x,q) : A$ for every labeled QBDTL formula $(i,x,q) : A$.[93] If $\models (i,x,q) : A$ is assumed, then $(i,x,q) : \neg A$ is unsatisfiable. A decision procedure can be defined, which is based on semantic tableaux with relation to the derivability of formulae in $\mathcal{N}(\text{QBDTL})$.[94] Step by step, it can be proven that if a global formula $@_j \neg A$ is unsatisfiable, then there is no Hintikka structure for $@_j \neg A$. If there is no Hintikka structure for $@_j \neg A$, then the root of a tableau for $@_j \neg A$ is "marked." If a node of a tableau is "marked," then it is inconsistent.

Until now, the formal system $\mathcal{N}(\text{QBDTL})$ of natural deduction is limited by its local perspective. In a branching-temporal logic, only

---

[92]Viganò *et al.* (2015, p. 21, Fig. 4).
[93]Viganò *et al.* (2015, pp. 26–33).
[94]The proof is inspired by Ben-Ari *et al.* (1983), Emerson and Halpern (1985).

the temporal evolution of local quantum states with respect to quantum systems of qubits can be modeled. Obviously, the quantum physical property of superpositions are grasped in the qubits ("agents") but not the other essential property of entanglement. Entanglement cannot be described by local states of quantum systems at distant locations but only as nonlocal states. Thus, the notion of synchronization of agents in $\mathcal{N}(\mathrm{QBDTL})$ is not sufficient because it relates to local agents which "communicate." Therefore, the quantum logic must be extended by a propositional symbol $entg_i$, which models the information that agent i is entangled and no local information about its quantum states is available.

In general, the growth of quantum technology needs secure quantum programming languages. In order to minimize risks of failures in programming, the verification of quantum programs is a great challenge of the future. A main challenge of quantum computing is the control of complex temporal processes with superposition and entanglement. At this point, temporal logic comes in to verify quantum programs.[95] The calculi of quantum temporal logic need completeness for reasoning in order to verify quantum programs. Furthermore, polynomial time algorithms are necessary to compute the reachability of computing processes and their average running time. In addition, the decidability of formulae in quantum temporal logic must be studied.

4.4. Societal Impact of Temporal Logic

4.4.1. *Generations of quantum technologies*[96]

We are currently living in the second generation of quantum technology, in which basic principles of quantum mechanics are specifically implemented in quantum-mechanical devices (Fig. 4.17). This includes the first prototypes of quantum computers, classical supercomputers with quantum simulation, quantum cryptography and quantum communication, quantum sensor technology, and quantum measurement technology (metrology). To solve a specific task, in 2019, a special quantum computer, called Sycamore, with 54 qubits

---

[95]Basin *et al.* (2011b).
[96]Mainzer (2020, Chapter 12).

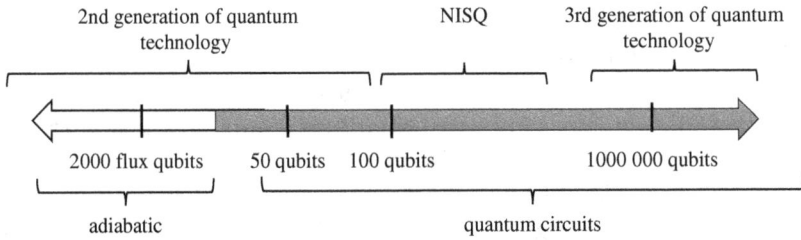

Fig. 4.17. Generations of quantum computers.

was able to demonstrate "supremacy" for the first time, i.e., the superiority of a quantum computer over a classical supercomputer. In the same year, a quantum computer with 20 qubits, IBM Q, had already been presented earlier, which is based, at least in principle, on the architecture of a universal multi-purpose computer. Both are not yet directly applicable for commercial use, but they were built by two of the largest IT corporations, IBM and Google, as milestones on the way to a universal quantum computer with supremacy. In the age of digitalization, the first classical digital universal computers were available in 1941 with Zuse's Z3 and in 1945–1946 with John von Neumann's ENIAC. Both opened up the development of a broad spectrum of digital technology at that time. With a universal quantum computer, the third generation of quantum technology would begin (Fig. 4.17).

Such a universal quantum computer, which could realize, for example, the Shor algorithm for the factorization of large numbers, would require millions of qubits. Because of the sensitivity of such quantum algorithms to noise, it would have to be able to carry out technically highly complex error corrections. Today, it has already been shown that a quantum computer with over 50 qubits can solve tasks at a speed that is not feasible for the fastest classical super-computers. This means that there are already quantum computers that achieve feats that classical computers are incapable of. On the other hand, they are not yet large enough to realize a fault-tolerant application of the known quantum algorithms.

This era between quantum computers with 50–100 qubits (e.g., Google's Sycamore) and the first universal quantum computer with 1,000,000 qubits and more was called "Noisy Intermediate-Scale Quantum" (NISQ) by the American computer scientist John

Preskill.[97] It is "noisy," i.e., determined by noise, because not enough qubits are available for error correction, and it is in the "intermediate-scale" because the qubit number is sufficient for supremacy proof but not yet sufficient for a universal quantum computer. The currently dawning NISQ era thus describes the transition from the second to the third generation of quantum technology (Fig. 4.17).

A look at the emergence of the first digital computers shows the enormous challenges that still need to be overcome before the first universal quantum computers with supremacy can be developed in the third generation of quantum technology. First, there is the hardware technology, where superconducting materials are emerging as a key component. The associated low-temperature physics seems most likely to be able to guarantee superpositions and entanglement with sufficient coherence time.

In the second generation of quantum technology, quantum computers are already in commercial use that are not based on universal quantum circuits along the lines of classical digital computers. This refers to adiabatic computers or quantum annealers, which are being used commercially by companies such as D-Wave. These include faster calculations of traffic flow and solutions to other optimization tasks.

As in the classical computer, the various levels of software build on the hardware. On the theoretical side, the most advanced is the mathematical algorithm theory based on the formalism of quantum mechanics (e.g., Shor, Grover). It must be related to models of quantum circuits, which in turn are based on the physical technology of the hardware. Up to the user, however, the various programming levels play a major role. There are initial approaches for commercially and scientifically usable programming environments, as the examples of Cirq by Google and Qiskit by IBM show.

In the second generation, quantum simulations are already highly developed. They facilitate, for example, the detection of material defects that have electromagnetic causes. In general, optical material properties can be determined in this way. In order to produce drugs efficiently and cost-effectively, complex molecular structures must be calculated. Quantum simulation will prove indispensable for

---

[97]Preskill (2018).

increasingly complex challenges in biotechnology. Finally, quantum computers and quantum simulation will make AI applications more efficient. Classical computers are clearly reaching their limits in the application of learning and search algorithms in the age of Big Data.

In the development of digitization, first came the universal digital computer in the 1940s and 1950s, and then digital communication technology until the digital Internet in the 1990s. In the development of quantum technologies, it is becoming apparent that quantum communication with fiber optic cables, satellite technology and/or superconductors can be commercially deployed at the same time or even earlier than the universal quantum computer. This is due to the already highly developed technology of fiber optic cables, satellites, and solid-state superconductors. However, quantum communication with fiber optic networks requires "amplifiers" (quantum repeaters) for long distances of over thousands of miles, the technical standards for which are still pending. Whether at least a proof of concept will be technically available in one or two years is currently still an open question. In any case, quantum repeaters will be indispensable building blocks for the commercialization of quantum communication with fiber optic networks. In contrast, satellite technology is expensive but already technically mature.

A central challenge is also to secure data communication in time with quantum cryptography before quantum computers break the classical codes. Quantum-mechanical encryption guarantees absolute security for data transfer — an indispensable prerequisite for, say, financial transactions, especially if digital currencies are used in the future. Prevention of manipulation of personal data, such as patient records in social networks, also requires better security, which quantum cryptography can provide. The age of digitalization has also seen a growing dependence of civilization on digital infrastructures. Cyberattacks can disrupt energy supply or logistical supply chains at any time, for example. In this case, too, quantum communication and quantum cryptography will be indispensable.

In summary, the development of the quantum computer also shows that in research, the path can already be the goal: Not only is the ultimate goal of the universal quantum computer revolutionary, but on the way there, a broad spectrum of new quantum technologies is being developed. For example, quantum-based measurement technology is opening up great economic potential, from navigation, geology, and Earth observation to medical diagnostics, industrial precision measurement technology, and military technology.

4.4.2. *From logical–mathematical foundations to machine learning and quantum technology*

The example of quantum technology illustrates how the application success of a technology can be deeply rooted in the logical foundations of science. In the beginning, there were fundamental discussions and thought experiments (e.g., entanglement), which were eventually translated into laboratory experiments. Ultimately, it is about the transition from experimental setups in the laboratory to robust, reliable and cost-effective devices. This requires supporting industrial engineering technology that makes the construction of these devices possible in the first place. This is referred to as "enabling engineering."[98]

Typical examples are quantum-compatible data acquisition and fast electronics for data processing with high time resolution, low dead times, optimized and parallelized data throughput, etc. with the corresponding software. From the first generation of quantum technology, lasers and detectors were added. Enabling technology also includes materials, components, and quantum technology devices and processes for positioning and implanting individual atoms, ions or molecules, vacuum technology, and optical precision assembly. A mature enabling technology influences the future of an innovation. In fact, the branch of development that can build on already highly developed technology with standards and norms often prevails. In the end, markets create a push-and-pull effect to steer technical developments in certain directions. Markets are no longer just about demonstrating feasibility in principle (e.g., the supremacy proof of Google's Sycamore) but about turnover and sales of commercial goods.

In this context, the development of the quantum computer is embedded in the development of quantum technologies, which in turn are part of the global trend of digitalization. The concept of the quantum computer is thus only a beacon of research development.

---

[98]Enabling technology in: *Business Dictionary*, http://www.businessdictionary.com/definition/enabling-technology.html#ixzz2lrYdBsg3.

The promise of the future "quantum computer," however, does not just mean a single device, such as Zuse's Z3 or von Neumann's ENIAC at the time, which will be available at a certain point in time, but the broad road of technological development that is already changing markets and civilization.

Therefore, the sociological phrase of "disruptive" technologies is misleading because it suggests a sudden event that only changes the world when it arrives. In fact, it is a broadly diversified development that appears rather continuously when the facts are known. Only to those who do not know or do not understand the basics and background, the development must appear abrupt. It therefore requires an in-depth logical–philosophical analysis and reflection to be able to strategically assess and evaluate such technology trends correctly.

Why logic and philosophy? Since antiquity, they have been the origin of the sciences, which have become increasingly specialized over the centuries. Even Newton, the founder of modern physics, had a chair in natural philosophy. Newton called his main work *Principia Mathematica Philosophia Naturalis*, i.e., mathematical principles or foundations of natural philosophy. Since the Aristotelian tradition, "natural philosophy" has encompassed what is now called natural research. Observation and logical analysis, even the first mathematical models, already existed before Galileo.

This also applies to logic, which was highly developed and specialized in the Aristotelian tradition until the late Middle Ages. Modern temporal logic builds directly on modal logic in the Aristotelian tradition. In mathematical logic and basic research at the beginning of the 20th century, however, logic also combined with mathematics and became a subfield of mathematics in proof theory. The methods of proof theory were instrumental in founding the algorithms of theoretical computer science.

In physics, especially meterology, observation and experiment have been systematically linked with mathematical models since the time of Galileo. But this was no more "disruptive" than the heliocentric model of Copernicus. Apart from the fact that Greek astronomers were already discussing the possibility of a heliocentric planetary model, the development of natural science was extremely diversified over a long period of time: The basic concepts and methods of natural science did not suddenly appear and were proclaimed like the

Ten Commandments by a Moses of science. Thomas Kuhn's distinction between "scientific revolutions" and "normal science" is at best woodcut-like and justified from a sociologizing and psychologizing perception. Anyone familiar with the development of logic in modern times knows that mathematical calculus and algorithm theory did not appear "all at once" in the 20th century but developed step by step from Leibniz via Bolzano, Boole, Peano, *et al.* to Frege and Russell.

The criticism of the superficial talk of "disruption" also applies to the appearance of quantum physics in the 20th century. Historically, quantum mechanics had to develop from familiar concepts, procedures, and ideas of classical physics, in which generations of physicists were educated until the beginning of the 20th century. However, classical mechanics does not correspond at all to our everyday ideas, as is always claimed. That bodies reduced to points should move in a vacuum on ideal mathematical curves is a highly abstract model that even today must seem completely abstract and detached from everyday experience to people before their first physics lessons. In fact, Aristotelian physics describes phenomenologically our everyday experience, according to which falling massive bodies sink to the ground in the air surrounding us in a highly complex way, as in a liquid. The model of classical mechanics was only easy to calculate once the mathematical calculations and algorithms were available, in which every A-level student must first be trained.

Those who take up physics studies, however, start with quantum mechanics in the first semester, learn the linear algebra and functional analysis of its mathematical calculi, and solve lots of practice problems. With this practice, familiarity and habit with the physical models behind the problems will set in. Classical mechanics now appears as a simplified model that only has approximate value in special cases. As a matter of course, the quantum world is perceived as the actual reality.

This is where a modern logic and philosophy of physics and technology comes in, with which the understanding of the concept of state in the quantum world, the meaning of superposition, entanglement and tunneling can be explained and deepened. Nothing about it is "unnatural," "puzzling," or "disruptive." In classical mechanics, too, we know the phenomenon of only calculating without understanding the conceptual connections. This becomes apparent at the latest when problems arise that deviate from the usual routine. Therefore, it would also be naïve to take the classical understandings of time and modalities in modal calculi and temporal calculi as "natural." Modern physics unmistakably shows that relativistic models and quantum physics are more "natural" than Newton's world, which is only a crude and simplistic approximation. A major aim of this book is to show the dependence of modal and temporal logic on physical models. In addition, the consideration of physical concepts of time is necessary to develop effective control procedures of time logic for practical applications of computer science. In practice, the end goal is verification and certification procedures of, for example, quantum programs to guarantee safe and responsible applications of quantum computers.

So, even in quantum technology, it is dangerous to lose sight of understanding the logical and quantum-physical basics. We then end up shambling along with previously successful routines and become "operationally blind" to new and unusual technical paths that the potential of the basic quantum-mechanical concepts opens up. Logic and philosophy thus aim at understanding the basics and thus contribute to a better understanding of the application possibilities. Philosophy of technology builds on this, dealing with the goals of technology design.

Logic and philosophy thus still ask today, as in the times of Aristotle, about the principles and foundations of our knowledge and its interdisciplinary connections in the various disciplines in order to be able to decide and act responsibly. Since antiquity, logic, the foundation of science and ethics, and philosophy have belonged together. Problem- and practice-oriented networking with the sciences makes up the special profile of philosophy in the globalized knowledge society.

The decisive factor here is that logic and philosophy are anchored in the minds of students, teachers, and professors of engineering, natural sciences, social sciences, and economics. Only through constant contact between research and teaching can logic and philosophy be prevented from taking off into the clouds of abstraction, becoming entrenched in the history of their disciplines, and losing contact with science. However, this is the only way to stimulate the necessary, fundamental discussion in the sciences on the part of philosophy. However, this also presupposes that philosophers have been trained in mathematics, computer science, physics, biology, sociology, and economics, for example, in order to be accepted as competent in these disciplines.

Philosophy must therefore by no means be reduced to its subsidiary subject, ethics, as is unfortunately often the fashion today. Aristotle wrote his little book, *Nicomachean Ethics*, on this subject, but it was an extensive compendium of the basics of physics and logic of his time. Only those who have understood the basics of science can enter innovative new territory in technology and, building on this basic knowledge, speak competently about ethical questions.[99]

The ethical challenges of technology can therefore only be met if the basics are understood. This is particularly evident in the topic of this book: As civilization becomes increasingly dependent on digitalization and AI, considerable security risks are raised. No one sees through or can control in detail what goes on, for example, in the nonlinear interactions of the neurons, synapses, and quantum bits of a neural network in machine learning or quantum computing. Therefore, these neural networks are "black boxes" for users and developers, raising fundamental questions of security, trust in technology, and responsibility.

Artificial neural networks are extremely effective in dealing with complex problems (real-world problems). What is missing, however, are specifications and standards for the security of their outputs. This is also true for quantum programming. For this, the black box of neural networks and quantum computing must be better understood, controlled and verified. The verification of neural networks and quantum programs, however, is a hard problem of knowledge: even the proof of simple properties turns out to be NP-complete within the framework of complexity theory. The reasons for this are the size of the practically applied networks (scaling) and the nonlinear activation functions of their neurons, which cannot be comprehended by humans on this scale and at this speed. Since neural networks are also subject to the dynamics of complex systems, they are often sensitive to small disturbances and changes in their inputs, which can

---

[99]Mainzer (2018).

build up to uncontrollable effects. The robustness and stability of the networks is therefore closely related to their security. In temporal logic, complex bifurcations of the information processes have to be analyzed, which also lead to nonlinear problems.

Different verifications can be specified for different classes of neural networks, which can be derived from various theories of logic and mathematics.[100] These include verification methods based on the satisfiability of formulae of Boolean propositional logic (SAT: satisfiability theories),[101] satisfiability of formulae of first-level predicate logic (SMT: satisfiability modulo theories), reduction to linear problems (MIP: mixed integer linear programming), and robustness of multi-layer perceptron networks (MLP: multi-layer perceptron). SAT and SMT verifications combine the classical AI (symbolic AI) of automated reasoning with machine learning.[102] MIP is based on the logic and algebra of linear programming. The robustness investigations of MLP apply insights from the theory of complex dynamical systems in machine learning.

However, with advances in quantum technology, there is a growing danger that quantum computers will crack the security codes on which financial, economic and defense systems are based. The verification methods of machine learning must therefore be extended to the possibilities of quantum computers and quantum communication. On the other hand, the fundamentals of quantum mechanics also open up an absolutely secure quantum cryptology for quantum communication networks. In the end, the goal must be the standardization (in Germany DIN standards and internationally, ISO standards) of AI, quantum technology, and quantum computers, as already exists in classical digitization technology. Temporal logic is an important tool here that needs to be developed further.

Modern computing is composed of a variety of different computing technologies and is reminiscent of an ecosystem. In a biological ecosystem, different plants and animals are adapted (adaptive) to or compete with each other. In the technical sciences and sociology of technology, this metaphor is used to describe a system of different

---

[100]Ehlers (2017).
[101]Biere *et al.* (2009).
[102]Pulina and Taccella (2012).

technologies that are adapted to or compete with each other, such as the ecosystem of computing.[103] It consists of adaptive or competing computing technologies for different purposes (e.g., quantum computing, supercomputers, workstations, internet, social media, platforms, PCs, laptops, smartphones). Quantum computing will be a sub-technology that is particularly used for the rapid processing of data-intensive and complex computing problems (e.g., particle physics, climate models, life sciences, medicine).

All these different computing technologies pose different problems for time logic to coordinate the information flows in time. To secure them, proof-theoretical justifications will be indispensable. However, it will not be sufficient, as in traditional temporal logic, to start from the everyday understanding of time in the formalisms of time logic. Since the 20th century, physics has fundamentally changed the modern understanding of time, which now also determines technology. In time logic, we are only just beginning to grasp these differences in proof theory, which needs interdisciplinary cooperation of proof theory, computer science, physics, technology, and philosophy. This book aims to motivate and contribute to this orientation in the foundational research of temporal logic.

---

[103]Stahl (2022).

References

J. F. Allen and P. Hayes. (1989). Moments and points in an interval-based temporal logic. *Computational Intelligence*, 5(4), 225–238.

Aristotle. (1956). *Die Lehrschriften* (translated by P. Gohlke), Vol. IV. 1, Paderborn.

A. Arnold. (1994). *Finite Transition Systems: Semantics of Communicating Systems*. Prentice Hall.

H. Andréka, J. X. Madarász and I. Németi. (2006). Logic of spacetime and relativity theory. In M. Aiello, I. Pratt-Hartmann and J. Van Benthem (eds.), *Handbook of Spatial Logics*. Dordrecht: Springer. https://doi.org/10.1007/978-1-4020-5587-4_11.

A. Aspect, P. Grangier and G. Roger. (1981). Experimental tests of realistic local theories via Bell's theorem. *Physical Review Letters*, 47, 460–463.

Augustine, *Confessiones* (English: Saint Augustine, Confessions (translated by H. Chadwick), Oxford's World's Classics, Oxford University Press: Oxford 2009).

C. Baier and J. P. Katoen. (2008). *Principles of Model Checking*. MIT Press: Cambridge MA.

S. Baratella and A. Masini. (2001). An approach to infinitary temporal proof theory. *Archive for Mathematical Logic*, 43, 965–990.

D. Basin, C. Caleiro, J. Ramos and L. Viganò. (2011a). Labelled tableaux for distributed temporal logic. *Journal of Logic and Computation*, 19(6), 1245–1279.

D. Basin, C. Caleiro, J. Ramos and L. Viganò. (2011b). Distributed temporal logic for the analysis of security protocol models. *Theoretical Computer Science*, 412(3), 4007–4043.

O. Becker. (1930). Zur Logik der Modalitäten. *Jahrbuch für Philosophie und phänomenologische Forschung* XI. Max Niemeyer Verlag: Halle.

J. S. Bell. (1964). On the Einstein Podolsky Rosen paradoxon. *Physics*, 1, 195–290.

N. Belnap. (1992). Branching space-time. *Synthese*, 92(3), 385–434.

J. van Benthem. (1983). *The Logic of Time*. Dordrecht: Kluwer Academic Publishers.

M. Ben-Ari, A. Pnueli and Z. Manna. (1981). The temporal logic of branching time. *POPL 1981*. New York: ACM Press, 164–176.

G. Benenti, G. Casati and G. Strini. (2008). *Principles of Quantum Computing and Information* Vol. I: Basic Concepts. Singapore: World Scientific.

E. W. Beth. (1955). Semantic entailment and formal derivability. *Nieuwe Reeks*, 18(13), 309–342.

A. Bertolino. (2000). Software testing research: Achievements, challenges, dreams. *Future of Software Engineering* (FOSE'07) 0-7695-2829-5/07 IEEE.

Y. Bertot and P. Castéran. (2004). *Interactive Theorem Proving and Program Development. Coq'Art: The Calculus of Inductive Constructions.* New York: Springer.

A. Biere, M. Heule, H. van Maaren and T. Walsh (eds.). (2009). *Handbook of Satisfiability*. Amsterdam: IOS Press.

L. Bolc and A. Szalas (eds.). (1995). *Time and Logic: A Computational Approach*. London: UCL Press.

J. Boudou, M. Diéguez and D. Fernández-Duque. (2017). A decidable intuitionistic temporal logic. In *26th EACSL Annual Conference on Computer Science Logic (CSL). Leibniz International Proceedings in Informatics (LIPIcs) 82*. Schloss Dagstuhl Leibniz-Zentrum für Informatik, 14:1–14:17.

P. Bourque and R. Dupuis. (2004). SWEBOK. *Guide to the Software Engineering Body of Knowledge*. Los Alamitos: IEEE Computer Society.

B. Borelli. (2008). *Proof-Analysis in Temporal Logic*. Ph.D. Dissertation, Università di Milano.

L. E. J. Brouwer. (1907). On the foundations of mathematics. In A. Heyting (ed.), *L. E. J. Brouwer Collected Works*, Vol. 1, pp. 11–101. North-Holland, Amsterdam, 1975.

L. E. J. Brouwer. (1927). Über Definitionsbereiche von Funktionen. *Mathematische Annalen*, 97, 60–75.

K. Brünnler and M. Lange. (2008). Cut-free sequent systems for temporal logic. *The Journal of Logic and Algebraic Programming*, 76, 216–225.

J. R. Büchi. (1962). On a decision method in restricted second order arithmetic. In *Congress in Logic, Method, and Philosophy of Science*. Redwood City: Stanford University Press, pp. 1–12.

J. Burgess and Y. Gurevich. (1985). The decision problem for linear temporal logic. *Notre Dame Journal of Formal Logic*, 26(2), 115–128.

S. Centrone and P. Minari. (2019). *Oskar Becker on Modalities*. Berlin: Logos Verlag.

S. Centrone and P. Minari. (2023). Becker's rule is not *Becker's rule*. In A. Klev and S. Rahman (eds.), *Festschrift for Göran Sundholm*. Berlin: Springer, to appear.

S. Chopoghloo and M. Monini. (2021). A strongly complete axiomatization of intuitionistic temporal logic. *Journal of Logic and Computation*, exab 041. https://doi.org/10.1093/logocom/exab041.

A. Church. (1941). *The Calculi of Lambda-Conversion*. Princeton: Library of America, (repr. New York 1965).

E. Clarke and E. A. Emerson. (1981). Design and synthesis of synchronization skeletons using branching time temporal logic. In *Workshop on Logics of Programs*. Berlin: Springer, pp. 52–71.

H. Comon-Lundh, M. Dauchet, R. Gilleron, C. Löding, F. Jacquemard, D. Lugiez, S. Tison and M. Tommasi. *Tree Automata Techniques and Applications*, online book http://tata.gforge.inria.fr.

G. De Giacomo and M. Y. Vardi. Linear temporal logic and linear dynamic logic on finite traces. In *IJCAI International Joint Conference on Artificial Intelligence*, pp. 854–860.

M. P. Deisenroth. (2013). A survey on policy search for robotics. *Foundations and Trends in Robotics*, 2(1), 1–142.

D. De Jongh. (2008). Beth's main results in mathematical logic. In J. van Benthem, P. van Ulsen and H. Visser (eds.), *Logic and Scientific Philosophy. An E.W. Beth Centenary Celebration*, p. 13.

S. Demri, V. Goranko and M. Lange. (2016). *Temporal Logics in Computer Science. Finite-State Systems*. Cambridge: Cambridge University Press.

H. Diels and W. Kranz. (1960/1961). *Die Fragmente der Vorsokratiker*. Berlin.

M. L. Dalla Chiara. (1986). Quantum logic. In *Handbook of Philosophical Logic*, III, Reidel, pp. 427–469.

A. Donzé and O. Maler. (2010). Robust satisfaction of temporal logic over real-valued signals. In *Lecture Notes in Computer Science*, Vol. 6246, pp. 92–106.

D. D'Souza and P. Shankar (eds.). (2012). *Modern Applications of Automata Theory*, HSc Research Monographs Series no. 2, Singapore: World Scientific.

M. Dummett. (1977). *Elements of Intuitionism*. Oxford: Clarendon Press.

R. Ehlers. (2017). Formal verification of piece-wise linear feed-forward neural networks. *ArXiv:1705.01320v3 [cs.LO]* August 2017, 2.

H.-D. Ehrich and C. Caleiro. (2000). Specifying communication in distributed information systems. *Acta Informatica*, 36, 591–616.

E. A. Emerson. (1990). Temporal and modal logics. In J. van Leewen (ed.), *Handbook of Theoretical Computer Science*. Vol. B: *Formal Models and Semantics*. Cambridge: MIT Press.

E. A. Emerson and E. C. Clarke. (1982). Using branching time temporal logic to synthesize synchronisation skeletons. *Science of Computer Programming*, 2, 241–266.

E. A. Emerson and J. Halpern. (1985). Decision procedures and expressiveness in the temporal logic of branching time. *Journal of Computer and Systems Science*, 30, 1–24.

E. A. Emerson and J. Y. Halpern. (1983). "Sometimes" and "not ever" revisited. On branching versus linear time temporal logic. *Journal of the Association for Computing Machinery*, 33, 151–178.

Enabling technology, in: *Business Dictionary* http://www.businessdictionary.com/definition/enabling-technology.html#ixzz2lrYdBsg3.

G. Etesi and I. Németi. (2002). Non-Turing computations via Malamenent-Hogarth space-times. *ArXiv:gr-qc/0104023v2-20* February 2002, 10–13.

H. Ferber. (1981). *Zenons Paradoxien der Bewegung und die Struktur von Raum und Zeit*. München: Franz Steiner Verlag.

M. Fischer and R. Ladner. (1979). Propositional dynamical logic of regular programs. *Journal of Computer and System Science*, 18, 194–211.

M. Fisher, D. M. Gabbay and L. Vila. (2005). *Handbook of Temporal Reasoning in Artificial Intelligence*. Amsterdam: Elsevier.

O. Friedmann, M. Latte and M. Lange. (2013). Satisfiability games for branching-time logics. *Logical Methods in Computer Science*, 9(4), 1–36.

D. M. Gabbay. (1996). *Labelled Deductive Systems*. Vol. 1. Oxford: Clarendon Press.

D. M. Gabbay, I. Hodkinson and M. Reynolds. (1994, 2000). *Temporal Logic: Mathematical Foundations and Computational Aspects*. Vol. 1, Oxford: Clarendon Press, Vol. 2, Oxford: Oxford University Press.

D. M. Gabbay, A. Pnueli, S. Shelah and J. Stavi. (1980). On the temporal analysis of fairness. *POL*, 1980, 163–173.

J. Gaintzarain, M. Hermo, P. Lucio, M. Navarro and F. Orejas. (2007). A cut-free and invariant-free sequent calculus for PLTL. In J. Duparc and T. Henzinger (eds.), *Proceedings of CSL 2997 LNCS* 4616, pp. 481–495.

A. P. Galton. (2008). Temporal logic. In E. N. Zalta (ed.), *The Standford Encyclopedia of Philosophy*, Fall 2008 Edition.

G. Gentzen. (1935). Untersuchungen über das logische Schließen. *Mathematische Zeitschrift*, 39, 176–210.

K. Gödel. (1933). Eine Interpretation des intuitionistischen Aussagekalküls. *Ergebnisse eines mathematischen Kolloquiums*, 2, 39–40.

K. Gödel. (1949). An example of a new type of cosmological solutions of Einstein's field equations of gravitation. *Review of Modern Physics*, 21, 447–450.

R. Goldblatt. (1980). Diodorean modality in Minkowski spacetime. *Studia Logic*, 39, 219–236.

R. Goré. (1999). Tableaux methods for modal and temporal logics. In *Handbook of Tableaux Methods*. Kluwer.

S. Gu, E. Holly, T. Lillicrap and S. Levine. (2016). Deep reinforcement learning for robotic manipulation with asynchronous off-policy updates. *ArXiv:1610.00633*.

M. Hanus. (1986). *Problem Solving in PROLOG*. Stuttgart: Vieweg+ Teubner.

S. W. Hawking and G. F. R. Ellis. (1973). *The Large Scale Structure of Space-Time*. Cambridge: Cambridge University Press.

S. W. Hawking and R. Penrose. (1970). The singularities of gravitational collapse and cosmology. *Proceeding of Royal Society (London)*, A 314, 529–548.

J. D. Hidary. (2019). *Quantum Computing: An Applied Approach*. Cham: Springer.

J. Hintikka. (1962). *Knowledge and Belief*. Ithaca, New York: Cornell University Press.

J. Hintikka and G. Sandu. (2009). Game-theoretical semantics. In K. Allan (ed.), *Concise Encyclopedia of Semantics*. Elsevier, pp. 341–343.

M. Hogarth. (1992). Does general relativity allow an observer to view an eternity in a finite time? *Foundations of Physics Letters*, 5, 173–181.

M. Homeister. (2018). *Quantum Computing verstehen*, 5th edn. Wiesbaden: Springer.

W. A. Howard. (1969). The formulae-as-types notion of construction. In J. P. Seldin and J. R. Hindley (eds.), *To H.B. Curry: Essays on Combinatory Logic, Lambda Calculus and Formalism*. Boston, MA: Academic Press, pp. 479–490.

N. Kamidie. (2006). An equivalence between sequent calculi for linear-time temporal logic. *Bulletin of the Section of the Logic*, 35, 167–193.

I. Kant. (1781, 1787). *Kritik der reinen Vernunft* (A, B). Engl. Translation: *Critique of Pure Reasoning* (1991), Cambridge: Cambridge University Press.

R. Kashima. (1994). Cut-free sequent calculi for some tense logics, *Studia Logica*, 53 (1994), 119–135.

H. Kawai. (1987). Sequential calculus for a first-oder infintary temporal logic. *Zeitschrift für mathematische Logik und Grundlagen der Mathematik*, 33, 423–432.

M. Keyl. (2002). Fundamentals of quantum information theory. *Physics Reports. A Review Section of Physics Letters*, 369, 431–548.

B. Kienzle (ed.). (1994). *Zustand und Ereignis*. Frankfurt: Suhrkamp Verlag.

S. C. Kleene. (1967). *Introduction to Metamathematics*, 5th edn. Amsterdam: North Holland.

C. Klüppelberg, D. Straub and I. Welpe. (2014). *Risk — A Multidisciplinary Introduction*. Berlin: Springer.

W. Knight. (2017). The dark secret at the heart of AI. *MIT Technological Revue*, 11, 1–22.

G. Kreisel. (1967). Informal rigour and completeness proofs. In I. Lakatos (ed.), *Problems in the Philosophy of Mathematics. Proceedings of the International Colloquium in the Philosophy of Science* 1965, Amsterdam, North-Holland, pp. 138–186.

S. Kripke. (1963). Semantical analysis of modal logic I. Normal modal propositional Calculi. *Zeitschrift für Mathematische Logik und Grundlagen der Mathematik*, 9, 67–96.

M. Lange and C. Stirling. (2000). Model checking games for CTL*. *Proc. Conf. on Temporal Logic. ICTL 2000*, 115–125.

M. Lange and C. Stirling. (2002). Model checking games for branching time logics. *Journal of Logic and Computation*, 12(4), 623–639.

T. Latvala, A. Biere, K. Heljanko and T. Junttila. (2004). Simple bounded LTL model checking. *Formal Methods in Computer-Aided Design*, 3312, LCNS, 186–200.

B. Lennarston and Q.-S. Jia. (2020). Reinforcement learning with temporal logic constraints. *ScienceDirect IFAC PapersOn Line 53-4*, 488–492.

S. Levine, P. Pastor, A. Krizhevsky and D. Quillen. Learning hand-eys coordination for robotic grasping with deep learning and large-scale data collection. *ArXiv 2016*. http://arxiv.org/abs/1603.021199v1.

X. Li, C.-I. Vasile and C. Belta. (March 2, 2017). Reinforcement Learning with temporal logic rewards. *ArXiv:1612.03471v2 [cs.AI]*.

K. Lorenz. (1968). Dialogspiele als semantische Grundlage von Logikkalkülen. *Archiv für mathematische Logik und Grundlagenforschung*, 11, 73–100.

P. Lorenzen. (1969). *Normative Logic and Ethics*. Bibliographisches Institut Mannheim.

K. Mainzer. (1977). Is the intuitionistic bar-induction a constructive principle? *Notre Dame Journal of Formal Logic*, 18(4), 583–588.

K. Mainzer. (1981). Logik des Beweisens. In *Vernunft, Handlung und Erfahrung* (Hrsg. O. Schwemmer), München, pp. 22–33, 133–134.

K. Mainzer. (1985). Der Intelligenzbegriff in wissenschaftstheoretischer und erkenntnistheoretischer Sicht. In W. Strombach, M. J. Tauber, and B. Reusch (eds.), *Der Intelligenzbegriff in den verschiedenen Wissenschaften.* Schriftenreihe der Österreichischen Computergesellschaft Bd. 28, Wien/München, pp. 41–56.

K. Mainzer. (1988). Philosophie und Geschichte von Raum und Zeit. In J. Audretsch and K. Mainzer (eds.), *Philosophie und Physik der Raum-Zeit,* Bibliographisches Institut Mannheim 2nd edn 1994, pp. 39–44 (chapter 3.3: Minkowski's spactime).

K. Mainzer. (1990). Knowledge-based systems. Remarks on philosophy of technology and Artificial Intelligence. *Journal for General Philosophy of Science,* 21, 47–74.

K. Mainzer. (2002). *The Little Book of Time.* New York: Copernicus Books (German: 5th edition C.H. Beck Verlag: München 2005).

K. Mainzer. (2015). Exploring complexity: Von Artificial Life und Artificial Intelligence zu cyberphysical systems. In S. Jeschke (ed.), *Exploring Cybernetics,* Berlin: Springer, pp. 123–150.

K. Mainzer. (2016). Towards a theory of intelligent complex systems: From symbolic AI to embodied and evolutionary AI. In V. C. Müller (ed.), *Fundamental Issues of Artificial Intelligence.* Switzerland: Springer, pp. 239–257.

K. Mainzer. (2017). From augmented reality to the Internet of Things: Paradigm shifts in digital innovation dynamics. In J. M. Ariso (ed.), *Augmented Reality.* Berlin Studies in Knowledge Research Vol. 11, Berlin: W. De Gruyter, pp. 25–40.

K. Mainzer. (2018a). *The Digital and the Real World. Computational Foundations of Mathematics, Science, Technology, and Philosophy.* Singapore: World Scientific, Chapter 6 (Intuitionistic Mathematics and Human Creativity).

K. Mainzer. (2018b). *Wie berechenbar ist unsere Welt? Herausforderungen für die Mathematik, Informatik und Philosophie im Zeitalter der Digitalisierung.* Wiesbaden: Springer Essentials.

K. Mainzer. (2019). *Artificial Intelligence. When Do Machines Take Over?* Berlin: Springer.

K. Mainzer. (2020). *Quantencomputer. Von der Quantenwelt zur Künstlichen Intelligenz.* Berlin: Springer.

K. Mainzer and L. Chua. (2013). *Local Activity Principle.* London: Imperial College Press.

K. Mainzer and R. Kahle. (2022). *Grenzen der KI: Theoretisch, Praktisch, Ethisch.* Berlin: Springer.

K. Mainzer, P. Schuster and H. Schwichtenberg (eds.). (2018). *Proof and Computation. Digitization in Mathematics, Computer Science, and Philosophy.* Singapore: World Scientific.

K. Mainzer, P. Schuster and H. Schwichtenberg (eds.). (2022). *Proof and Computation II. From Proof Theory and Univalent Mathematics to Program Extraction and Verification.* Singapore: World Scientific.

F. Melia. (2000). The heart of the Milky Way. *American Scientist*, 346–353.

P. Mittelstaedt. (1986). *Sprache und Realität in der modernen Physik.* Mannheim: Bibliographisches Institut Mannheim.

M. Mithun. (2001). *The Languages of Native North America.* Cambridge: Cambridge University Press.

J. M. Mooij, D. Janzing and B. Schölkopf. (2013). From ordinary differential equations to structural causal models: The deterministic case. In *Proceedings of the 29th Annual Conference on Uncertainty in Artificial Intelligence* (UAI), pp. 440–448.

D. E. Muller, A. Saoudi and P. E. Schupp. (1988). Weak alternating automata give a simple explanation of why most temporal and dynamic logics are decidable in exponential time. *LICS 1988 IEEE*, 422–427.

S. Negri. (2005). Proof analysis in modal logic. *Journal of Philosophical Logic*, 34, 5017–5544.

S. Negri and B. Borelli. (2010). On the finitization of Priorean linear time. In G. D'Agostino *et al.* (eds.), *New Essays in Logic and Philosophy of Science*. London: College Publications, pp. 1–15.

H. Nishimura. (1980). A study of some tense logics by Gentzen's sequential method. *Publications of the Research Institute for Mathematical Sciences (RIMS). Kyoto University*, 16, 343–353.

P. Ohrstrom and P. Hasle. (1995). *Temporal Logic. From Ancient Ideas to Artificial Intelligence*. Dordrecht: Kluver Academic Publisher.

B. Paech. (1988). Gentzen systems for propositional temporal logics. In E. Börger, H. Kleine Brüning and M. Richter (eds.), *CSL 88. Proceedings of the 2nd Workshop on Computer Science Logic*. Berlin: Springer, pp. 240–253.

J. Pearl. (2009). *Causality: Models, Reasoning, and Inference*, 2nd edn. New York: Cambridge University Press.

R. Penrose. (1965). Gravitational collapse and space-time singularities. *Physical Review Letter*, 14, 57–59.

J. Peters, D. Janzing and B. Schölkopf. (2017). *Elements of Inference. Foundations and Learning Algorithms*. Cambridge, MA: MIT Press.

J. F. Phillips. (1998). A note on the modal and temporal logics for n-dimensional spacetime. *Notre Dame Journal of Formal Logic*, 39(4), 545–553.

A. Pnueli. (1977). The temporal logic of programs. In *Proceedings of the 18th IEEE Symposium on Foundations of Computer Science*, pp. 46–67.

F. Poggiolesi. (2011). *Gentzen Calculi for Modal Propositional Logic*. Dordrecht: Springer, p. 211.

V. Pratt. (1979). A practical decision method for propositional dynamics logic. In *10th Annual ACM Symposium on the Theory of Computing ACM*, pp. 326–337.

V. Pratt. (1980). A near optimal method for reasoning about actions. *Journal of Computer and System Sciences*, 20, 231–254.

J. Preskill. (2018). Quantum computing in the NISQ era and beyond. *arXiv:1801.00862v3 [quant-ph]*.

A. N. Prior. (1957). *Time and Modality*. Oxford: Oxford University Press.

A. N. Prior. (1967). *Past, Present and Future*. Oxford: Oxford University Press.

L. Pulina and A. Taccella. (2012). Challenging SMT solvers to verify neural networks. *AI Communications*, 25, 117–135.

F. Puppe. (1988). *Einführung in Expertensysteme*. Berlin: Springer.

N. Rescher and A. Urquhart. (1971). *Temporal Logic*. Berlin: Springer.

M. Reynolds. (2001). An axiomatization of full computation tree logic. *Journal of Symbolic Logic*, 66(3), 1011–1057.

M. Reynolds. (2007). A tableau for bundled CTL*. *Journal of Logic and Computation*, 17(1), 117–132.

M. Reynolds. (2009). A tableau for CTL*. *FM 2009 Lecture Notes in Computer Science Vol. 5850*. Berlin: Springer.

M. Reynolds. (2011). A tableau-based decision procedure for CTL*. *Formal Aspects of Computing*, 23(6), 739–779.

M. Reynolds. (2013). A faster tableau for CTL*. *GandALF 2013*, 50–63.

J. A. Robinson. (1965). A machine-oriented logic based on the resolution principle. *Journal of the Association for Computing Machinery*, 12, 23–41.

S. Russell and P. Norvig. (2004). *Künstliche Intelligenz: Ein moderner Ansatz*. München: Pearson Studium.

A. Santelli (ed.). (2022). *Ockhamism and Philosophy of Time. Semantic and Metaphysical Issues Concerning Future Contingents*. Berlin: Springer.

I. Shapirovsky and V. Shehtmann. (2003). Chronological future modality in Minkowski spacetime. *Advances in Modal Logic*, 4, 437–459.

G. Sundholm. (2014). Constructive recursive functions, Church's thesis, and Brouwer's theory of the creative subject: Afterthoughts on a Parisian Joint session. *Dubucs Bourdeau*, 2014, 1–35.

R. Sutton and A. Barto. (1998). *Reinforcement-Learning: An Introduction*. Cambridge, MA: A Bradford Book.

M. Takano. (2018). A semantical analysis of cut-free calculi for modal logics. *Reports on Mathematical Logic*, 53, 43–65.

J. Tretmans and E. Brinksma. (2003). TorX: Automated model-based testing. In A. Hartman and K. Dussa-Zieger (eds.), *Proceedings of the First European Conference on Model-Driven Software Engineering*.

A. S. Troelstra. (1968). The theory of choice sequences. In B. van Rootselaar and J. F. Staal (eds.). *Logic, Methodology, and Philosophy of Science* 3, Amsterdam, North-Holland, pp. 201–223.

S. L. Uckelman and J. Uckelman. (2007). Modal and temporal logics for abstract space-time structures. *Studies in History and Philosophy of Science. Part B: Studies in History and Philosophy of Modern Physics*, 38(3), 673–681.

J. van Benthem, P. van Ulsen and H. Visser (eds.). (2008). *Logic and Scientific Philosophy. An E.W. Beth Centenary Celebration.* Amsterdam: Evert Willem Beth Foundation.

M. Y. Vardi and P. Wolper. (1986). Automata-theoretic approach to automatic program verification. *LICS 1986 IEEE*, 322–331.

M. Y. Vardi and P. Wolper. (1994). Reasoning about infinite computation. *Information and Computation*, 115, 1–37.

L. Viganò. (2000). *Labelled Non-Classical Logics.* New York: Kluwer Academic Publishers.

L. Viganò, M. Volpe and M. Zorzi. (2015). A branching distributed temporal logic for reasoning about quantum state transformations. *hal-01213511.*

L. Vila. (1994). A survey on temporal reasoning in Artificial Intelligence. *AI Communications*, 7, 4–28.

G. H. von Wright. (1974). Determinismus, Wahrheit und Zeitlichkeit. Ein Beitrag zum Problem der zukünftigen kontingenten Wahrheiten. *Studia Leibnitiana*, 6, 161–178.

W. Wahlster and C. Winterhalter (eds.). (2020). *German Standardization Roadmap on Artificial Intelligence.* DIN Berlin. https://www.dke.de/de/arbeitsfelder/core-safety/normungsroadmap-ki.

H. Wansing. (2002). Sequent systems for modal logics. In D. Gabbay and F. Guenther (eds.), *Handbook of Philosophical Logic*, 2nd edn, Vol. 8, S. Klüver, pp. 61–145.

H. Weber. (1893). Leopold Kronecker. In Deutsche Mathematische Vereinigung (Hrsg.), *Jahresbericht der Deutschen Mathematiker-Vereinigung*, Band 2, Reimer, pp. 5–31.

S. Weinberg. (1972). *Gravitation and Cosmology. Principles and Applications of the General Theory of Relativity.* New York: John Wiley & Sons.

G. Winkel and M. Nielsen. (1995). Event structures. In *Handbook of Logic in Computer Science* IV, Oxford: Oxford University Press.

P. Wolper. (1983). Temporal logic can be more expressive. *Information and Control*, 56, 72–99.

P. Wolper. (1985). The tableau method for temporal logic: An overview. *Logique et Analyse*, 110–111, 119–136.

Index